Philosophy of Interdisciplinarity

Interdisciplinarity is a hallmark of contemporary knowledge production. This book introduces a *Philosophy of Interdisciplinarity* at the intersection of science, society and sustainability. In light of the ambivalence of the technosciences and the challenge of sustainable development in the Anthropocene, this engaged philosophy provides a novel critical perspective on interdisciplinarity in science policy and research practice. It draws upon the original spirit of interdisciplinarity as an environmentalist concept and advocates an essential change in human-nature relations. The author utilizes the rich tradition of philosophy for case study analysis and develops a framework to disentangle the various forms of inter- and transdisciplinarity. *Philosophy of Interdisciplinarity* offers a foundation for a critical-reflexive program of interdisciplinarity conducive to a sustainable future for our knowledge society and contributes to fields such as sustainability science, social ecology, environmental ethics, technology assessment, complex systems, philosophy of nature, and philosophy of science. It injects a fresh way of thinking on interdisciplinarity – and supports researchers as well as science policy makers, university managers, and academic administrators in critical-reflexive knowledge production for sustainable development.

Jan Cornelius Schmidt is Professor of Philosophy at Darmstadt University of Applied Sciences.

History and Philosophy of Technoscience
Edited by Alfred Nordmann

Philosophy of Interdisciplinarity

Studies in Science, Society and Sustainability

Jan Cornelius Schmidt

Routledge
Taylor & Francis Group

LONDON AND NEW YORK

First published 2022
by Routledge
2 Park Square, Milton Park, Abingdon, Oxon OX14 4RN, UK

and by Routledge
605 Third Avenue, New York, NY 10158, USA

*Routledge is an imprint of the Taylor & Francis Group, an
informa business.*

British Library Cataloguing-in-Publication Data
A catalogue record for this book is available from the British
Library

Library of Congress Cataloging-in-Publication Data
A catalog record has been requested for this book

ISBN: 978-1-138-23007-1 (hbk)
ISBN: 978-1-032-11846-8 (pbk)
ISBN: 978-1-315-38710-9 (ebk)

DOI: 10.4324/9781315387109

Typeset in Sabon
by SPi Technologies India Pvt Ltd (Straive)

In memory of my father
Dr. Wolf-Rüdiger Schmidt

Contents

Acknowledgements

I am deeply grateful to the series editor, Alfred Nordmann, for encouraging me to finalize this study and for publishing my book as part of the book series "History and Philosophy of Technoscience". I sincerely thank Susan Keller for her very helpful editing suggestions, careful proofreading, and invaluable linguistic support. I also thank Stephen Poole for providing editorial support and Stefan Gammel for proofreading and providing help with the index. Furthermore, I am indebted to Peter Euler, Robert Frodeman, Armin Grunwald, Michael Hoffmann, Britt Holbrook, Thomas Jahn, Wolfgang Krohn, Wolfgang Liebert, Stephan Lingner, Sabine Maasen, Klaus Mainzer, Alfred Nordmann, Theres Paulsen, Christian Pohl, and Jakob Zinsstag for immensely inspiring and engaging discussions on inter- and transdisciplinarity over the past 20 years. My sincere thanks also go to Darmstadt University of Applied Sciences and in particular to the Vice-President for Research and Sustainability, Nicole Saenger, for much appreciated financial support. Finally, I am very thankful for the help and patience of the senior editors at Routledge, Bob Langham and Max Novick.

1 Introduction

What does the philosophy of interdisciplinarity offer?

A critical attitude towards knowledge production and technoscientific progress was once a cornerstone of inter- and transdisciplinarity. The focus was on pressing environmental problems, the challenges of global change, and side effects of technological advancement. The role of science and of the entire academic system in causing the non-sustainable state of our lifeworld came under scrutiny. The mindset of modernity, including the power and authority of science to dominate our way of framing and conceptualizing nature, was questioned. Occasionally, interdisciplinarity—along with its cognate transdisciplinarity—was deemed a fundamental challenge to the academy within late-modern knowledge societies.

Today, however, interdisciplinarity has lost its critical momentum and its original spirit. It has been reduced to a trendy, tame, and toothless notion. The term is omnipresent in science, technology, and economy as well as in society and higher education—fuelling the rhetoric of knowledge politics in our late-modern knowledge societies. It is increasingly being used as a synonym for application-oriented research and, in particular, for the commercialization of the university in neoliberal times.[1] Most researchers, politicians, and economists regard interdisciplinarity as a positive factor and value in itself which merits and compels support.[2] The National Academy of Sciences (2005), for instance, has declared "facilitating interdisciplinary research" one of its chief goals. Interdisciplinarity is seen as a kind of a panacea capable of curing pathologies of academic and entrepreneurial knowledge production (cf. Frodeman 2014). Clearly, no one is willing or able to resist its pull. Who would not subscribe to interdisciplinarity?

Since Erich Jantsch (1970, 1972) introduced the umbrella term to a larger audience in the early 1970s, "interdisciplinarity" has experienced an impressive career of wider recognition.[3] It is attributed to a plethora of research, innovation, or education programs. A highly esteemed initiative of the National Science Foundation regards the development of "Converging Technologies"—namely nanotechnology, biotechnology, information technology, and cognitive science (NBIC)—as being necessarily interdisciplinary (Roco and Bainbridge 2002). Interdisciplinarity serves as a fashionable label for innovation programs, for economics, for business and management studies, for military research, or for programmatic strategies of institutions

DOI: 10.4324/9781315387109-1

of higher education.[4] Even basic research programs seem to be highly inter-disciplinary; for example, physical cosmology has become established as an interdisciplinary effort at the intersections of physics, mathematics, chemistry, geology, and computer science.

In addition, interdisciplinarity is ascribed to issue-driven research at the interfaces between science, technology, and society: technology assessment, global change studies, sustainability science, risk management, and social ecological research.[5] Some scholars have identified the emergence of joint problem solving among science, technology, and society, which often is referred to as transdisciplinarity.[6] The particular attribution of transdisciplinarity and, more recently, the labels of transformation or transition[7] seem to be strong new threads in the discourse on interdisciplinarity. Other scholars prefer cognates, such as multi-, pluri-, cross-, meta-, or infra-disciplinarity.

Generally, inter- and transdisciplinarity are seen as central notions for the diagnosis of a current shift in the mode of scientific knowledge production, most popularly characterized by terms such as mode-2 science, post-normal science, post-paradigmatic science, post-academic science, technoscience, problem-oriented research, post-disciplinarity, triple helix research, trans-formative research, transition science, and citizen science.[8]

This book does not solely investigate whether the programmatic catch-words carry any distinctive content and any *differentia specifica*. In addition, it aims at a different view of the things attached to the concepts.

Interdisciplinarity—here provisionally used as an umbrella term to include the popular cognate notion of transdisciplinarity—is virtually ubiquitous, but what ends and purposes does it serve? What does interdisciplinarity entail, and what is its significance?

Going beyond the hype of the catchwords, by shifting the focus onto what is at stake when it comes to our natural environment and our socio-eco-logical life-world in the age of the *Anthropocene*, a philosophic approach rooted in rich cultural tradition could strengthen critical voices amid the recent hype. These voices—questioning the function and role of science within society and advocating a different way of viewing nature and a new relation of humans to nature—do not constitute the mainstream today.[9] However, a review of the term's history reveals that, right from its origin, interdisciplinarity has always carried an inherently normative dimension interlaced with a critical momentum: Interdisciplinarity has served as a syn-onym, *first*, for a cultural and social critique of knowledge production and specifically for a critique of the sciences within society and their view of nature and, *second*, for the engagement of the academic system in bettering socio-ecological *praxis and poiesis* of the human–nature relation. The cri-tique was loudly and broadly articulated during the student revolt and amid the first wave of recognition of environmental problems in the late 1960s and early 1970s—when the term "interdisciplinarity" became prominent.[10]

Beyond today's strong and narrow focus on the involvement of extra-scien-tific actors, lay people, and other stakeholders,[11] inter- and transdisciplinarity are, according to Jantsch (1972, 107/100f), the key notions for "renew[ing] the education and innovation system."[12] For Jantsch, there is an obvious need for a renewal: He observes "degrading side-effects of technology on the systems of human life," particularly on the "natural environment." Moreover, "there does not seem to be any alternative if a rational, one might even say, an ecological approach to science and technology is mandatory, as indeed it has to be so considered in the present situation" (ibid., 111/120). What is strongly required, Jantsch states, is a different way of shaping and designing the "joint systems of society and technology" (ibid., 119). In this vein, Jantsch argues for a new agenda and a different orientation of science and society—an orientation that, today, is frequently labelled with the umbrella term "sustainable development" (WCED 1987). Recognizing deficits in the academy and research practices, Jantsch sees inter- and transdisciplinarity as a lever and as a momentum for a self-renewal of the academic system (ibid., 100). The *Philosophy of Interdisciplinarity* set forth in this book concurs with Jantsch's intention. However, his specific line of argumentation, which was formulated about 50 years ago in times of planning optimism and control euphoria, will not be further pursued.

In contrast to the term's ubiquitous usage today, the original idea(l) of interdisciplinarity was more distinctive and richer in content when Jantsch first used it. Interdisciplinarity was intertwined with concerns about the environment and loss of biodiversity; it was interlinked with an in-depth consideration of the human–nature relations and with the predominantly scientific view of nature and the environment; it was interlaced with a call for scientists (and the sciences) to take responsibility and accountability for society at large; and it was interwoven with the recognition of the inherent ambivalence of science (and technology) throughout the emerging knowledge societies: Scientific knowledge is a doubled-edged sword. Science has contributed to inducing the environmental crisis, on the one hand, but at the same time it is deemed key to finding possible solutions. The ecological crisis is, in fact, a cultural and societal crisis.

The cultural background of the crisis, entailing a disciplinary blindness regarding the complex human–nature relations, has been seen as intricately interwoven with the sciences and the academy.[13] In this vein, John Ziman (1987) raises doubts as to whether disciplinary "knowing everything about nothing" is a reliable ground for shaping the future of our knowledge societies: Interdisciplinarity is the response to the overspecialization of the academy and to the isolated silos created by disciplinary methods, languages, and frameworks. Certainly, the shift towards interdisciplinarity is undeniably a good move—and we are on the way to establishing interdisciplinary practice on various levels in the academy and beyond. On the other hand, is such an overall positive appreciation of interdisciplinarity a plausible position? Is it the adequate approach for meeting the present-day environmental challenges facing the sciences in society?

This book asks whether such a position is delusive or, at least, one-sided. If we are aiming at a sustainable orientation of the academy embedded in late-modern societies, we need further differentiations and a different take on inter- and transdisciplinarity. The basic distinction set out and advanced throughout this book is that made between an instrumentalist or strategic account of interdisciplinarity, on the one hand, and a critical-reflexive or communicative kind of interdisciplinarity, on the other hand—a differentiation also roughly undertaken by others.[14] Other frequently suggested distinctions such as the involvement—or non-involvement—of lay people and stakeholders in the process of knowledge production are seen to be of minor relevance.

Present-day inter- or transdisciplinary practice, as well as the scholarly debate about it, is dominated by an instrumentalist or strategic viewpoint:[15] Methods and management procedures are prioritized. Recipes and organizational guidelines to facilitate interdisciplinarity are being developed (National Academies of Science 2005; Newell 2001; Newell 2013; Repko and Szostak 2017). Talk of "integration methods" is prominent.[16] In order to foster procedures of interdisciplinary integration, a new field, termed "science of team science," in which interdisciplinarity is seen as a collaboration and management challenge, has been established. But to what ends and what purposes does interdisciplinarity serve? If interdisciplinarity is the answer or the remedy, then what is the question? More specifically, what is the problem to be tackled?

The mantra that we need *more* interdisciplinarity to respond to and to overcome the overspecialization of the academy is no doubt more than convincing. However, this true but trivial statement hinders a deeper and more thorough reflection on (and justification of) the ends and purposes of interdisciplinary engagement. Symptomatically, instrumentalists and strategic actors tend to remain silent on normative aspects of science and society: They hardly reflect explicitly (as part of their scientific action and their research practice) on goals, problems, values, underlying convictions, or institutional structures.[17] Instead, instrumentalists accept given purposes as the point of departure for starting their research and moving ahead. Although they are aware that values can originate from different sources— from curiosity and epistemic contexts or from non-epistemic contexts such as ethical, societal, economic, or even religious contexts (cp. Machamer and Wolters 2004; Gethmann et al. 2015)—and that pursuing certain value-related goals might be preferable to striving for others, instrumentalists do not consider it part of their interdisciplinary endeavour to explicitly examine or justify goals.[18] The reflection on and (possibly) the revision of values or goals (or both) seem beyond the scope of their professional inclinations and even competences as scientists. Central to this means-centred view is its strict solution orientation: The feasibility of finding an ultimate and benign solution is associated here with the term "problem." Instrumentalists advocate *solutionism*,[19] namely that solutions exist and, furthermore, that these solutions ensure an ultimate elimination of the problem.[20] Interdisciplinary projects, it appears, develop means and instruments to reach solutions.

Accordingly, instrumentalists approve the traditional separation between science and society or between the academy and politics. In doing so, they perpetuate the chimeric view that science—including interdisciplinary research—has to be value-free in its epistemic core or that it can, at least, be purified and cleansed from its intertwinement with the societal domain.[21] Consequently, they share the ideal of means/ends rationality and rely on traditional action theories that draw a dividing line between means, delivered by value-free research, and ends, provided by the social, political, or ethical sphere.[22] Interdisciplinarity shows up as an organizational and management challenge—regardless of the aims of the various projects in the diverse fields such as quantum cosmology, synthetic bio-materials, nano-optics, personalized medicine, military research, nuclear reactor design, innovation studies, risk management, sustainability studies, or geoengineering.

The critical-reflexive approach, as pursued in this book, does not deny the existence of deficits in the organizational means, in the management tools or institutional procedures to foster and facilitate interdisciplinary practice—that is not at all the case. But it takes the focus to a deeper level of reflection. Deficits are perceived in the underlying rationality and in the dominant, means-focused production of knowledge, which is intertwined with how nature and the societal relations to nature are viewed. This is where the critical-reflexive position sets in: It scrutinizes knowledge production in our late-modern societies, focusing in particular on what is acknowledged as scientific knowledge. It urges scientists to rethink science—a helpful catchphrase advanced by Nowotny et al. (2001) and others (Nicolescu 2002, 2008). The critical-reflexive position argues that there is a need to rethink thinking—and this book presents examples: We should seek alternative ways of framing nature and different configurations of knowledge that are not related to the Baconian power mode of action and not as reliant on the modern dichotomist mindset as the Cartesian or Kantian viewpoints. We cannot tackle problems with the same kind of thinking we used when we created them. This post-Baconian/Cartesian/Kantian perspective can be linked to a different kind of accountability and responsibility of scientists with regard to society at large and, specifically, with regard to the natural environment.

Following this line of argumentation, critical-reflexive interdisciplinarity can be seen as a synonym for self-awareness, self-critique, and self-reflexivity: briefly, of self-enlightenment of the academy in neoliberal times. It is linked with the search for alternative views of nature[23] and for a different mindset—in other words, with the quest for other directions of scientific progress. Clearly, such a critical-reflexive approach draws on the critical theory and cultural critique of the Frankfurt School (Horkheimer 1972; Horkheimer and Adorno 1972), and it shares many dimensions with Habermas's concept of communicative action (Habermas 1984, 1993).[24]

Although both positions—the strategic-instrumentalist and the critical-reflexive—exist in interdisciplinary practice, the former is much more common. Besides being present within inner-academic interdisciplinarity, it determines most transdisciplinary technology-centred practices as well as science policy and innovation programs.[25] A further aspect, and one which this book finds important to consider from a critical perspective, is that it also features prominently in a subset of transdisciplinary engagement, namely in problem-oriented interdisciplinarity—a field that also includes an extra- or trans-scientific perspective, but in a specific way, as we will see.[26] The following examples illustrate the breadth of this standpoint: Klein et al. (2001) characterize transdisciplinarity as "joint problem solving among science and society." Jochen Jaeger and Martin Scheringer (1998) call it a "problem-related form of science." Gotthard Bechmann and Günter Frederichs (1996) locate it as "problem-oriented research in between public policy and science" (see also Newell 2013; Repko et al. 2017). Britt Holbrook (2013, 1867) argues (from a critical angle) that "transdisciplinarity refers to the integration of one or more academic disciplines with extra-academic perspectives on a common (and usually a real-world, as opposed to a merely academic) problem." Wolf Krohn (2010, 33) takes this parlance further when he states: "Any research field or research project that addresses real-world problems is considered to be essentially interdisciplinary." In sum, the view shared among many scholars is that the task is to provide solutions to extra-scientific, real-world problems—based on the assumption that they are in principle solvable.

This book shows that such an instrumentalist approach is, indeed, a step forward because it conceptualizes science as being not only curiosity-driven but also trans-epistemic–oriented *and* shapeable by stakeholders; epistemic and non-epistemic values are interlaced. However, given the current situation of the global change crisis, it seems necessary to question whether the instrumentalist account of interdisciplinarity is strong enough to cope with the above-mentioned challenges. For example, in a different field, the instrumentalist approach to solving problems such as growing traffic jams could be to build bigger highways or more garages. However, this solution fails to address the continuous production of new problems and the origin of the problems (e.g., more cars induce more traffic jams) caused by people's misguided habits and the worldwide misconception of mobility. There is a great need to complement "problem solving" with "problem prevention"—based on a "problem radar" to anticipate and to detect socio-ecological issues before they even emerge. The instrumentalist focus is too narrow to be able to address the wider cultural fundament of global change problems. Since the instrumentalist approach does not question the internal configuration and logic of knowledge production in the disciplines, it seems far too tame and weak to enhance self-reflexivity and responsibility among the academic communities and their disciplines—or to address the root of continuous problem production.

A closer examination of the relationship between the instrumentalist view and the critical-reflexive approach reveals that the distinction is not absolute; we find a kind of 'dialectic' relation, which is not disjunctive or antagonistic in the sense that it requires us to subscribe to either one or the other understanding. Rather, the instrumentalist position can be taken as a necessary first step towards a critical-reflexive type of interdisciplinarity.

In short, the latter complements the former. Not only is the critical-reflexive understanding based on instrumentalist considerations but, moreover, it can be regarded as instrumentalist on a higher or deeper level—since it aims to achieve a much bigger impact on a sustainable development of science, technology, and society. In light of the foregoing, the *Philosophy of Interdisciplinarity* proposed here aims to deepen and strengthen the instrumentality of inter- and transdisciplinarity—beyond the shortcomings of instrumentalism with its reductionist focus on means and its non-recognition of the role of science in neoliberal societies: How can the impact of interdisciplinarity and transdisciplinarity—and thus the contribution of the academy to sustainable development—be improved in such a fundamental way that it reaches the underlying cultural, cognitive, and intellectual basis of human action in late-modern societies? In this regard, the critical-reflexive approach in interdisciplinarity can be seen as *meta-instrumentalist* or as *deeply instrumentalist*.

Advocating a critical-reflexive concept entails going beyond the analytic tradition that dominated the philosophy of science throughout the 20th century, at least in the Anglo-American world. The analytic approach in philosophy was certainly decisive in facilitating philosophic inquiry based on criteria of clarity, distinctness, and precision. There is, however, a flipside. It has also narrowed philosophy's self-understanding and self-conceptualization; in particular, it has led to a retreat of philosophy from the public arena to the ivory tower, as Robert Frodeman (2014) argues (cp. also Frodeman 2010; Frodeman and Briggle 2016). The core question today is how to make philosophy more "relevant and pertinent" to the world we live in and at the same time avoid the shortcomings and ineffectiveness of mere application-oriented approaches.[27] Interdisciplinarity thus can be perceived as a fundamental challenge to philosophy itself. It urges us to rethink what philosophy is, as an academic discipline, and what it ought to be.

In essence, the envisioned *Philosophy of Interdisciplinarity* is not intended or to be regarded as a new subdiscipline of philosophy in the same way as, for example, philosophy of science, epistemology, neuro-philosophy, or bio-ethics, which presuppose domain-restricted approaches or ontologies. Such "hyphen-philosophies" or "philosophies of" merely increase the total amount of fragmented knowledge (cp. also Frodeman 2014). In contrast, the *Philosophy of Interdisciplinarity* facilitates a larger perspective of things and a wider approach. Many philosophical subdisciplines are involved: from ethics, anthropology, and the history of philosophy, through social, cultural, and political philosophy, to philosophy of science and technology. However, that is not all: *Philosophy of Interdisciplinarity* goes even further by reflecting on and synthesizing the insights of other disciplines—and, if necessary, knowledge of stakeholders and life-world actors. The *Philosophy of Interdisciplinarity* is thus interdisciplinary *and* genuinely philosophical: In comparison with the disciplinary mainstream of 20th-century philosophy with its subdisciplines, its reductionist approaches and regional ontologies (cp. Frodeman 2014), the *Philosophy of Interdisciplinarity* can be characterized as truly interdisciplinary. Furthermore, it is genuinely philosophic because it is based on the rich and colourful intellectual tradition of philosophy that addresses fundamental metaphysical questions and develops frameworks of orientation. In other words, the *Philosophy of Interdisciplinarity* aims to (re)open the academic discipline of philosophy towards other disciplines and, beyond that, to society at large. It resonates with an interdisciplinary-oriented philosophy and therefore could also be called *interdisciplinary philosophy*.

Such philosophic engagement in the general discourse on interdisciplinarity is not reserved exclusively to professional (disciplinary) philosophers. Philosophic thinking also takes place outside the domains of institutionalized philosophy. Embracing this broader understanding of philosophy as an art of inquiring, questioning, and critique is a prerequisite for any endeavour that can be called *Philosophy of Interdisciplinarity*. It is essential since interdisciplinarity is undoubtedly an interdisciplinary topic. This circular nature of interdisciplinarity needs to be considered. None of the academic disciplines institutionalized in the 20th or early 21st century can claim to be the one and only appropriate authority or to have an exclusive grasp on interdisciplinarity. The word "philosophy" used in the title of this book encompasses a wider understanding of philosophy that traces back to the intellectual tradition of non-disciplinary philosophy prevailing throughout pre-19th-century cultural history.

Such a critical-reflexive concept of interdisciplinarity, which aims to inject a fresh wave of rethinking thinking and to cultivate a critical mindset in the academy, does, indeed, not stand in isolation. It comes close to what Julie Thompson Klein (2010, 22f) calls, from a somewhat different angle, "critical interdisciplinarity." Liora Salter and Alison Hearn (1996) consider "critical interdisciplinarity" a "fundamental challenge to the disciplines" and to disciplinarily organized knowledge production. Referring to the Frankfurt School of critical theory, Peter Euler (1999, 299) discloses a "critical attitude attached to the notion of interdisciplinarity" that "challenges the seamless web of science and technology on the one hand, and global capitalism on the other hand."[28] Diana Hummel, Thomas Jahn, and others (2017) conceptualize "social ecology as a critical, transdisciplinary scientific research program for studies of societal relations to nature": By drawing on elements of the Frankfurt School and criticizing what they see as the "dualistic mindset of modernity," they claim that "transdisciplinarity allows for a reflection of the social contexts of scientific knowledge production" (ibid., 15). Going back to the 1980s, Jürgen Mittelstraß (1987) argued in his seminal publication that interdisciplinarity is "a reminder of the original ideal of science" to provide a "normative orientation knowledge" that goes beyond strategic and means-focused "disposition knowledge":[29] Orientation knowledge in this sense can be provided only if research succeeds in synthesizing diverse approaches and disciplinary propositions—and in this process re-installs the unity of rationality.

In a similar vein, but without such a strong emphasis on rationality, Steve Fuller (2010) advocates a concept named "deviant interdisciplinarity." This kind of interdisciplinarity differs from what he terms "normal interdisciplinarity" and features—in the tradition of philosophical thinking in the wake of German Idealism—a critical and synthetic orientation; its focus is to reflect on and, where appropriate, to shape the various underlying worldviews of nature and of societal relations to nature.[30] A critical line of thought is also pursued by Basarab Nicolescu (2002, 2008) in his programmatic "Manifesto of Transdisciplinarity," in which he advances criticism of the dualist mindset of modern knowledge by referencing quantum physics and its holistic concept of nature: Transdisciplinarity offers novel ways of perceiving and thinking which transgress modern dichotomies, in particular the "split between science and meaning, between subject and object" (Nicolescu 2008, 13).

Other thinkers may also support the basic premise of this book. Michael O'Rourke's and Stephen J. Crowley's "Toolbox Project," which might convey an instrumental connotation at first glance, in essence fosters critical reflexivity among scholars and researchers by means of specific, well-chosen "philosophical interventions" in the various cognitive processes underlying interdisciplinary communication (O'Rourke and Crowley 2013). O'Rourke and Crowley develop a strong approach known as "philosophic intervention research" and "engaged philosophy" and in doing so promote an "understanding [of] philosophy in both its critical and facilitative roles" with respect to the various societal challenges attributed to interdisciplinary research (ibid., 1939).

Robert Frodeman, most notably, takes a different and in-depth critical-reflexive approach in his editorial introduction to the *Oxford Handbook of Interdisciplinarity* (Frodeman 2010) and in his book advancing *A Theory of Interdisciplinarity* (Frodeman 2014). According to Frodeman, interdisciplinarity questions the fundamental premises of what is referred to as the knowledge society—namely that the pursuit of knowledge is always good and beneficial. He critiques the curiosity-driven ideal of the infinite growth of scientific knowledge. Frodeman (2010, xxx) advocates an inherently critical element in interdisciplinarity, for instance, when he addresses the question of what is "pertinent knowledge" and whether "knowledge is pertinent at all."[31] Interdisciplinarity sets out to "dediscipline philosophy" and to conceptualize "philosophy as interdisciplinary." Frodeman's inspiring seminal work and his take on inter- and transdisciplinarity are very much in line with the objective of this book, in particular when he speaks of "critical interdisciplinarity" (Frodeman and Mitcham 2007).

There are two central questions that the *Philosophy of Interdisciplinarity* needs to address. *First* of all, what is the essence of "interdisciplinarity" and cognates such as "transdisciplinarity"? How can we characterize the phenomena and the history of interdisciplinarity? *Second*, how should "interdisciplinarity," and in particular "transdisciplinarity" and "problem-oriented interdisciplinarity," be understood in a normative sense? In other words, how should we interpret the notion of interdisciplinarity in order to value and foster self-reflexivity, self-awareness, and self-critique—and which meaning enables us to facilitate responsibility and accountability within the academy for society at large? How can we reduce the rate of continuous problem production of the joint systems of science, technology, and society?

These questions are not solely of academic interest. How interdisciplinarity is framed and defined has implications for knowledge politics, public research policy, and research practices.[32] Interdisciplinarity therefore can be regarded as a signifier of the centrality of a debate on the future of knowledge. An objective of this book is to provide an understanding that will enable us not only to draw clear lines in the jungle of definitions of interdisciplinarity but also to inject a more critical-reflexive approach into recent knowledge politics—and to further substantiate the pathway to environmentally oriented sustainable development.

The *Philosophy of Interdisciplinarity*, as proposed in this book, offers a framework for such an engaged program. The task it seeks to accomplish is to not only keep up with the evolving area of inter- and transdisciplinary practice but also to advance critical-reflexive potential concerning the future of science, society, and sustainability in neoliberal times. The argumentation of the *Philosophy of Interdisciplinarity* is built around seven areas of focus:

- Taking stock and providing clarification of different types of interdisciplinarity and transdisciplinarity;
- Addressing and assessing the knowledge and research politics of interdisciplinarity;
- Tracing the historical roots of the dominant instrumental account of interdisciplinarity and linking interdisciplinarity to the discourse on technosciences;
- Explicating and reflecting on societal problems and opening science to society, giving substance to transdisciplinarity in particular;
- Engaging with the grand challenges and reinforcing the approaches of environmental ethics;
- Searching for alternative directions in the sciences and conceptualizing a different view of nature that stops the continuous production of environmental problems;
- Advancing and facilitating a prospective technoscience assessment to trigger a change in knowledge and research politics.

This book is structured around these seven areas of focus that the *Philosophy of Interdisciplinarity* intends to promulgate.

More specifically, the *Philosophy of Interdisciplinarity* takes its point of departure from the recognition that both "interdisciplinarity" and "transdisciplinarity" are poorly understood notions requiring further philosophic engagement. Chapter 2, entitled "Philosophy and Plurality," reveals different interpretations of the generic term "interdisciplinarity" and includes current popular concepts of "transdisciplinarity." To provide conceptual clarification, key epistemological questions behind "interdisciplinarity" are addressed. The driving idea is to develop a philosophical fundament for a critique of innumerable usages of the term.

Moreover, philosophers can go beyond an analytical classification and give more substance to the evolving discourse. On scrutiny, the term "interdisciplinarity" can also be perceived as a political buzzword in knowledge policy which is charged with strong promises and high expectations with regard to innovation. The *Philosophy of Interdisciplinarity* aims to provide a critique of recent technoscientific programs: Chapter 3 is dedicated to "Politics and Research Programs." It addresses the "politics of interdisciplinarity." Although an analysis of this kind is by necessity concept-centred, the basic intention has a practical dimension: to disclose the guiding values and ideals behind research programs that pre-determine our societal futures and to enable a broader public awareness and political discourse. This chapter reviews the assumptions of one of the most influential technoscientific research programs of the last 20 years, which presses for a strong instrumental kind of object-centred interdisciplinarity.

Viewing interdisciplinarity in this way is by no means novel. Chapter 4—"History and Technoscience: Tracing the Historical Roots"—offers further insights into the instrumentalist account of interdisciplinarity by tracing it back to the beginning of the modern age and to Francis Bacon. The historical perspective also encompasses a review of the present-day discussion surrounding the popular label "technoscience." Technoscience and object-oriented interdisciplinarity turn out to be twins.

In subsequent chapters, it is argued that a more critical-reflexive perspective is possible—and indeed needed in the discourse and practice of interdisciplinarity. Chapter 5—"Society and Societal Problems: Conceptualizing Problem-oriented Inter- and Transdisciplinarity"—opens science to society at large. Focusing attention on societal problems might seem indicative of an instrumentalist orientation. However, this is not the whole story since a critical-reflexive account is also central to this notion of interdisciplinarity.

A short interlude chapter addresses some shortcomings of the instrumentalist view of interdisciplinarity. Then, Chapter 6—"Ethics and the Environment: Engaging with Grand Environmental Challenges"—moves to an ontological focus and metaphysical reflection intersecting with ethics, sciences, and philosophy of nature.

An understanding of critical-reflexive interdisciplinarity as an ethical and environmentalist concept is developed and applied to two fields of endeavour: to the development of novel and different ways of viewing nature based on alternative concepts of science and scientific knowledge, on the one hand, and to knowledge politics and technoscience assessment, on the other hand. Chapter 7, headed "Nature and the Sciences: In Search of Alternative Concepts of Nature and Science," develops an intra-scientific critique of the mainstream sciences and seeks to come up with different, science-based concepts of nature. Chapter 8 on "Technology and the Future" advances a critical-reflexive and prospective-oriented approach in Technoscience Assessment and develops a critique of synthetic biology.

Although this book approaches inter- and transdisciplinarity from a theoretical perspective, the intention is essentially a practical one: to reach, in the classic sense of philosophic thinking, the *praxis* of the sciences and the academy in *and for* society. Motivated by the challenge of the global change crises in the age of the *Anthropocene*, the aim is to support the recent momentum towards a transformation (a) of the research enterprise and the academy and (b) of the way nature is viewed. The *Philosophy of Interdisciplinarity* stands in the environmentalist tradition. It subscribes to the thinking of the environmental philosopher Hans Jonas, who states that the recent socio-ecological crisis should be viewed as a cultural and societal one that questions our ways of thinking and perceiving—briefly, our mindset and attitude in our societal relations to nature.

In other words, the call for inter- and transdisciplinarity has to be associated with the ideal of sustainable development and socio-ecological transformation on a deeper level; it entails a cultural critique of knowledge production since the knowledge production is, in fact, strongly interlaced with the ongoing problem production. Such a perspective counters the mainstream rhetoric through which "interdisciplinarity" has become an omnipresent buzzword supporting the commercialization of the academy in neoliberal times. Interdisciplinarity holds a hidden critical, and transformative, potential; it carries momentum to raise fundamental questions in order to open up avenues and stimulate a broader engagement of the academy towards a sustainable future. This mindset is today mostly labelled "transdisciplinarity," although not all transdisciplinary approaches are societally centred and critically reflexive (see the next chapter for an in-depth discussion). In sum, the general discourse on "interdisciplinarity" can be seen as a signifier of what Ulrich Beck (1992) calls "reflexive modernization."

Concisely, this book argues that the buzzwords "inter- and transdisciplinarity" are not only indicative of a wound in the established culture of knowledge production, which can be cured by some minor (management) re-adjustments. It asserts that inter- and transdisciplinarity signify a thorn digging in the heart of the academy and the sciences. The relevance and pertinence of traditional academic knowledge for our late-modern society are at issue. In addition, and complementarily to the plausible instrumental and strategic demand for new means and new kinds of management capabilities, inter- and transdisciplinarity call for deeper reflection on and a revision of the values and norms, ontological backgrounds, and metaphysical convictions governing our framing, perceiving, and understanding of nature and of science in societies. In sum, inter- and transdisciplinarity challenge the modern dualistic mindset and the way in which nature is perceived and conceptualized.

Put succinctly, inter- and transdisciplinarity urge us to rethink our thinking and to reframe our framing of nature; also, they compel us to reflect on and change our practice in science, research, and education. Those are the main messages conveyed by this book, building on the intellectual spirit and original momentum of both philosophy and interdisciplinarity. A philosophic endeavour of this kind cannot be apathetic or indifferent about the world. It is not a value-free enterprise. Rather, it is concerned with the world's state of affairs: with the grand environmental challenges and global change issues facing late-modern societies under the pressure of the capitalist market. The *Philosophy of Interdisciplinarity*, in this light, is part of what is called *transformative* or *transition research*: research *for* (and not only *on*) transformation (Hummel et al. 2017; Krohn et al. 2017). It not only reflects on an epochal break in the culture of knowledge production (that goes beyond questions of stakeholder involvement) but also aims to foster and facilitate a new critical-reflexive practice in (and of) the academy.

Notes

1 A neoliberal dimension in the discourse on inter- and transdisciplinarity is observed by Maasen (2010) and Frodeman (2014).
2 There are some exceptions; see, for instance, Jacobs and Frickel (2009), Jacobs (2013), and Graff (2015).
3 Although the term's origin is a matter of dispute, most scholars cite Jantsch (cf. Klein 1990, 1996), whereas Nicolescu (2002) credits Piaget (cp. also Kockelmans 1979). For the history of "transdisciplinarity," see also Bernstein (2015).
4 Examples include innovation studies (Fagerberg et al. 2005). For a critical take, see Cozzens and Gieryn (1990).
5 These fields of interdisciplinary engagement are vigorously promoted, for example, by Kates et al. (2001), Decker (2001, 2004), Decker and Grunwald (2001), Becker (2002), Norton (2005), Becker and Jahn (2006), and Lingner (2015).
6 This is a commonly held view among those who are engaged in the discourse. See, for example, Klein et al. (2001), Pohl and Hirsch Hadorn (2007), Hirsch Hadorn et al. (2008), or Bogner et al. (2010); for a consideration of different meanings of the terms, see Wickson et al. (2006), and for a distinctly different approach, see the work by Mittelstraß (1987). A thought-provoking explication of transdisciplinary research is given by Krohn et al. (2017).
7 See the debate on transformative research: Schneidewind and Singer-Brodowski (2013), Strohschneider (2014), and Grunwald (2015).
8 The literature on these topics is overwhelming, particularly in sociology of science and in science and technology studies (STS). Here is a short list of publications: Gibbons et al. (1994), Funtowicz and Ravetz (1993), Elzinga (1995), Böhme et al. (1983), Ziman (2000), Bammé (2004), Haraway (1991), Latour (1987), Nordmann (2004), Nordmann et al. (2011), Schmidt (2004), Ihde and Selinger (2003), Chubin et al. (1986), de Bie (1970), Bechmann and Frederichs (1996), Becker and Jahn (2006), Becker (2002), Norton (2005), Holbrook (2015), Etzkowitz and Leydesdorff (1998), Irwin (1997), Hirsch Hadorn et al. (2008), Bogner et al. (2010), Kastenhofer (2010), Moran (2010), Riesch and Potter (2014), Szostak et al. (2016), and the seminal edition by Kocka (1987). From a more philosophic perspective, see Kötter and Balsiger (1999), Hubig (2001), Schmidt (2003), Balsinger (2005), Grunwald and Schmidt (2005), Schmidt (2005), Schmidt (2008b), Frodeman (2010, 2014), Jungert et al. (2010), Holbrook (2010, 2013), Wechsler and Hurst (2011), Gethmann et al. (2015), Szostak (2015), Krohn et al. (2017), and Hummel et al. (2017).

9 Whereas some of the points raised above have been addressed over the past decades by social scientists, notably in the field of science and technology studies (STS), philosophers, surprisingly, have so far stayed on the sidelines for most of the time.

10 There are some exceptions. For example, with regard to their concept of transdisciplinarity in social ecological research, see Hummel et al. (2017), who draw on critical theory of the Frankfurt School.

11 For a critique, see Mittelstraß (2018) and Jaeger and Scheringer (2018).

12 An early critical perspective on the education system is pursued in Kockelmans (1979).

13 This thesis is defended by environmentalists such as Jonas (1984), Holsten (1988), or Hummel et al. (2017).

14 See, exemplarily, Klein (2010, 5/22f), Frodeman and Mitcham (2007), and Salter and Hearn (1996). Moreover, O'Rourke and Crowley (2013) and Holbrook (2013) draw explicitly on communication in this regard. By distinguishing between an instrumentalist and a critical-reflexive/communicative view, the approach pursued in this book also leans on Habermas's "Theory of Communicative Action" and his distinction between communicative and strategic rationality (Habermas 1984). The notion of "critical interdisciplinarity" dates back at least to Gusdorf (1977).

15 This shortcoming is also perpetuated by the National Academies of the US: "A deeper understanding of these processes [of interdisciplinary knowledge production] will further enhance the prospects for creation and management of successful IDR [interdisciplinary research] programs" (National Academies 2005, 3).

16 For a critique, see Holbrook (2013), Frodeman (2010, xxxii), Frodeman (2014), Fuller (2017) and, from a different angle, Nicolescu (2002, 2008).

17 The term "critical interdisciplinarity" is also elaborated by Frodeman and Mitcham (2007). As Klein (2010, 23) underlines, "Critical ID [= the critical-reflexive type of ID] interrogates the dominant structure of knowledge and education with the aim of transforming them, raising questions of value and purpose silent in Instrumental ID."

18 Fuller (2017), for instance, identifies an ambivalent "military-industrial route to interdisciplinarity" and reveals a "Janus-faced character of […] interdisciplinarity."

19 This notion is taken from Morozov (2013).

20 In contrast, since in many and the most urgent cases (e.g., global change) an ultimate solution does not exist, the critical-reflexive approach is not primarily centred on solutions. More fundamentally, the critical-reflexive approach deals with problems on a deeper level, in particular with the origin and emergence of problems.

21 For instance, the Roco–Bainbridge report on converging technologies sees interdisciplinarity from such an instrumentalist perspective (Roco and Bainbridge 2002, xiii/76): "The scientific and engineering communities should create new means of interdisciplinary training and communication." Yet "our nation needs to formulate a new interdisciplinary, inter-science, and systems-wide collaborative model based on converging NBIC technologies in order to move forward to create productive and efficient change." This view is dominant among the public as well as in academia.

22 By this, they subscribe to the similar split between facts and values; for an analysis, see Kincaid et al. (2007).

23 This includes societal relations to nature (Jahn et al. 2012; Hummel et al. 2017).

24 Discourse-ethical considerations are central to the critical-reflexive understanding of interdisciplinarity. Some elements going in this direction are advocated by Holbrook (2013), who identifies a "Habermas–Klein" view of interdisciplinarity, and by O'Rourke and Crowley (2013).

25 See, for example, the *Oxford Handbook of Innovation* (Fagerberg et al. 2005).

26 "Transdisciplinarity" and "problem-oriented interdisciplinarity" differ in a certain way, as will be discussed in the next chapter. The notion of "problem" will also be discussed further on.

27 My translation (J.C.S.). The approach advocated in this book shares many aspects with Frodeman (2010, xxix f, 2014) and contrasts with what Gibbons et al. (1994) labelled "mode-2-research." Clearly, a "mode-2-philosophy" is not envisioned here.

28 Euler (1999) refers to Mikosch's (1993, 55f) "critical theory of interdisciplinarity."

29 My translation (J.C.S.). Mittelstraß (1987, 156) maintains that "transdisciplinarity [...] does not leave the disciplinary things and the academy untouched. Instead, transdisciplinarity carries a momentum that could [and should] have a broader impact on the disciplines and the science system at large"; see also Hummel et al. (2017, 10), Jahn et al. (2012), and Jahn (2013).

30 According to Fuller (2010), philosophy in the "normal" mode plays an auxiliary role beside other disciplines and accepts the division of organized inquiry into disciplines. Fuller argues that philosophy should transcend disciplinary knowledge and actively promote an encompassing and synthetic understanding of reality. He refers to the long intellectual tradition of philosophic thinking and, in particular, to natural philosophy. The notion of "worldview" echoes the German term "Weltbild" and, in particular, encompasses underlying concepts of how we see and perceive the world (nature, humans, society, sciences ...).

31 Frodeman (2010, xxxi) urges a reorientation of philosophy in its mindset, in its attitude and approach to the world, as "interdisciplinarity represents the resurgence of interest in a larger view of things." Philosophy should engage in the world, Frodeman argues, and to do so it needs to transform towards becoming a "field philosophy."

32 In addition, there seems to be a need for a "political philosophy of science" (Rouse 1987).

2 Philosophy and plurality

Providing a classification and clarification of interdisciplinarity

Hot topic

Since the early 1970s and a path-breaking congress of the Organisation for Economic Co-operation and Development (OECD) in Paris, the need for a conceptual clarification of the term "interdisciplinarity," along with cognates such as "transdisciplinarity," has become obvious. Inter- and transdisciplinarity are in vogue in science, society, and economy. At the same time, both terms remain misty and unclear.

The vagueness might have posed a particular challenge to philosophers of science and analytic philosophers. Across all traditions and schools, they share the belief that a clear, distinct, and rigorous terminology is essential for knowledge generation and for communication.[1] Yet, although such a clarification might have become a canonical task, philosophers seem to feel uncomfortable addressing such a hot topic. This book intends to challenge their reluctance—and to support the scholarly and public debate on interdisciplinarity.[2] The aim in this second chapter is to provide a philosophical fundament—rooted in the tradition of philosophy—for a deeper clarification of the term. On that basis, we will later be able to develop a specific critical point of view.

Philosophy proves to be a rich resource for untangling the notion of interdisciplinarity. Referring to well-established distinctions in the philosophy of science, this chapter argues for a unity in plurality by examining four different types of interdisciplinarity in public and scientific discourses, namely interdisciplinarity with regard to objects (ontology); knowledge, concepts, and theories (epistemology); methods and heuristics (methodology); and pressing societal problems and issues. The philosophical framework of the four types or, we could say, the four framings of interdisciplinarity will be best illustrated by research programs that are prominently labelled "interdisciplinary." As will be discussed, it is striking that different philosophical traditions can be related to these four types. In fact, the different traditions determine which understanding of interdisciplinarity is favoured—and which types of interdisciplinarity are regarded as plausible and which not. Conversely, interdisciplinarity can serve as an excellent thematic focus for an introduction to philosophy of science—or, more precisely and

DOI: 10.4324/9781315387109-2

provocatively, to a "political philosophy of science" (Rouse 1987) or a "philosophy of science policy" (Frodeman and Mitcham 2004). The chapter reveals that a *minimal philosophy of science* constitutes an indispensable cornerstone of the *Philosophy of Interdisciplinarity*.[3]

Richness of the tradition

Philosophers seem to doubt whether the recent popularity of interdisciplinarity is justified. They are sceptical whether the label itself refers to a noticeably new mode of research. Seen in this light, interdisciplinarity appears to be merely a public, political, or ideological term that is part of a popular rhetoric and little more than a kind of parlance. This perception has fuelled scepticism about its value. Since philosophers claim to focus only on semantically relevant terms, interdisciplinarity is not regarded as a serious field of inquiry.

An additional reason for the reluctance of philosophers might be the mere fact that the phenomena associated with interdisciplinarity seem too complex, too heterogeneous, too dynamic, and too contextual to be accessible for philosophy, particularly for the philosophy of science. Interdisciplinary practice discourages a philosophical approach, as it appears to be non-universal, non-theoretical, context-specific, case-restricted, strongly value-laden, and often driven by non-epistemic values.[4] Expressed more provocatively, the limits of philosophy of science, notably in the analytic tradition, result in a reduced interest in interdisciplinarity among philosophers of science.

Despite the general reluctance, we find on closer examination that philosophy provides a rich framework for addressing interdisciplinarity. Although the word itself is not in philosophy's core vocabulary, the associated phenomena and topics are well known and hotly debated among philosophers. We find productive lines of thought in domains typically labelled monism, dualism, pluralism, inter-theoretic relations, holism, unification, and reduction. The ontological, epistemological, and methodological issues involved have occupied and challenged philosophic thinking since ancient times. In addition, new fields of philosophic engagement are paving the way towards a *Philosophy of Interdisciplinarity*. These include, for example, the history, sociology, and ethics of science; philosophy of technoscience; social epistemology; and political philosophy of science—all of which represent novel, vibrant, and exciting areas of philosophic inquiry. Some lines of thought to illustrate the richness of the tradition for our endeavour will be briefly outlined below.

But before doing so, let us explore what constitutes the field of inquiry of such a philosophic approach. A central goal of interdisciplinary practice is to bridge different disciplines, which leads to a certain level of integration and even to a synthesis or unification. Interdisciplinarity seems to be strongly needed in order to compensate for what has been lost over time: Although the functional differentiation and separation into disciplines have undeniably contributed to the impressive success of scientific explanation of the world and to the overall historical advancement of science—as seen, for instance, in quantum physics, cosmology, evolutionary theory, and synthetic or systems biology—there is a flipside. A patchwork of knowledge fragments, methods, and objects can be observed today. Diversity and even an overall disunity of sciences become apparent (Galison and Stump 1996). The academy appears to be fragmented or, worse, fractured into silos of disciplinary specialization: "knowing everything about nothing" (Ziman 1987).

Interdisciplinarity counteracts this development. It is regarded as a corrective or compensatory effort to regain a common way of looking at the world or even to achieve unity within the patchwork of disciplinary knowledge. The quest for such an integrative approach is by no means novel. Although the need for integration did not become apparent until the 20th century, in the period when differentiation, specialization, and fragmentation were at their strongest, integration has been an overall aim of academic inquiry since the ancient Greeks and, notably, since Plato. Leibniz, for instance, later renewed the goal of finding a common denominator and a synthesis of the world's fundamental knowledge with his ideal of a *mathesis universalis*. Traditional natural philosophy in the 18th and early 19th century sought unity in the diversity of the novel scientific insights on nature, the cosmos, and man. According to Hegel and others in the period of German Idealism, the truth has to be associated with the whole and not primarily with specialized, splintered, disciplinary insights.

Although the lines of arguments may have changed during the 20th century, the pursuit of integration of knowledge across disciplines remains as topical as ever. Most interestingly, and complementarily to the historical process of fragmentation, we can identify movements towards integration in many areas of the sciences themselves—forming a core element of synthesis or, stronger, of (inner-)disciplinary reduction. For instance, physicists have been and continue to be very successful at integrating and unifying different theories. They search for a "theory of everything" with the aim of bringing four fundamental forces or theories into a coherent body of a grand unified theory. Philosophers of science, standing in the tradition of the Vienna Circle and the Unity of Science movement of the 1920s and 1930s, have greatly valued the approach taken by physicists for being paradigmatic for the progress of science and scientific explanation. Strong forms of integration can be seen as a reduction[5]—meaning the dissolution of one theory into another such that the latter is then acknowledged as the more fundamental one. Accordingly, some advocates of interdisciplinarity turn out to be reductionists, too.

In contrast to the strong positions on integration and unification, which are not very common among interdisciplinarians, there are weaker and much more moderate positions to be found in the discourse surrounding interdisciplinarity. These focus on the "particulate unity of the empirical object," as Helmut Schelsky (1961) puts it.[6] The weaker positions presuppose a local, contextual, and provisional unity with regard to one object or domain instead of an overall unity throughout the entire world. They aim to address "the complexity, the totality, and the unity of one single object," as Ursula Hübenthal (1991) argues.[7] Often, these weaker positions on unity are developed from a pragmatic problem-oriented or real-world perspective with the goal of focusing on societally relevant objects or problems, which are so wicked, complex, and interrelated that a disciplinary approach is usually not feasible. Interdisciplinarity is regarded as a tool for tackling these complex issues. Methodological considerations for technology assessment, sustainability science, and social-ecological research have been developed along this line of thought (Decker 2001; Norton 2005). A certain local monism concerning objects and problems seems to be consistent with a global pluralism concerning methods, concepts, propositions, theories, and worldviews.

Whereas this weaker (second) position on unity often shows up in connection with integrating epistemic (intra-scientific) and non-epistemic (external, extra-scientific) values and is issue-driven (e.g., by global climate-change problems), the strong (first) position is an internal one. The internal–external distinction reflects the common parlance pertaining to interdisciplinarity and has given rise to the notion of "transdisciplinarity," which typically refers to the second position. This distinction, including the demarcation between science and society, touches on hot topics of present-day philosophy, such as the value dimension of science, the amalgamation of truth and power, questions concerning the legitimacy and authority of science in society at large, or the governance and shaping of science and research. Such topics are vigorously debated in new philosophical directions, namely in *social epistemology* (Fuller 2002).

In general, the two positions—the strong and the weak—share a positive view on the possibility of interdisciplinarity and its efficiency. Other positions, or thought traditions, are more pessimistic. A most prominent example of the latter might be Neo-Kantianism, including what has become known as the philosophy of culture. In the late 19th century, scholars such as Heinrich Rickert, Wilhelm Dilthey, and Wilhelm Windelband developed philosophical approaches underscoring the differences and unbridgeable gaps between natural science and the humanities—including the liberal arts, history, cultural studies ("Kulturwissenschaften"), and many fields of social science. These scholars favour demarcations and suggest several criteria to justify and defend the humanities, cultural studies, and social sciences as a distinct form of epistemic enterprise. In support of the battle against the growing dominance of the natural sciences in defining what epistemic knowledge and academic expertise are or ought to be, they also point to Immanuel Kant's thinking and his work on the "conflict of the faculties" from 1798. From today's perspective, Kant's work can be seen as a milestone in reflecting on the dissonance and unbridgeable gaps between the different disciplines.

A parallel stream of discourse deserves to be mentioned because it also shows the philosophic nature of the issues involved in interdisciplinarity. Since the late 1950s and a seminal essay by Charles Percy Snow, the term

"Two Cultures" has enjoyed an impressive growth in popularity (Snow 2001). Snow coined the term to characterize the very different mindsets, convictions, habits, socializations, and worldviews of natural scientists, on the one hand, and those of scholars from the humanities, liberal arts, and cultural studies, on the other. For interdisciplinarians, Snow's observation appears rather frustrating. Bridging the two-culture gap hardly seems possible at all.[8] Snow's thesis could also be derived from Thomas Kuhn's concept of paradigm or from Ludwik Fleck's idea of thought styles published about two decades earlier—although Kuhn and Fleck refer mainly to (intra-) disciplinarity (Kuhn 1970; Fleck 1979): If communication between communities that subscribe to different paradigms or thought styles is barely possible within one discipline, the same holds to an even greater extent for the communication between different disciplines. It can therefore be maintained that Kuhn and Fleck come to an even more pessimistic assessment than Snow. For the critical analysis undertaken here, it suffices to note that philosophy, particularly the philosophy of science, and the domain of social epistemology that became established from Kuhn onwards have much to offer when dealing with issues of interdisciplinarity.

Another confirmation of the pessimistic stance about interdisciplinary collaboration across the two-culture gap arose in the mid to late 1990s following an "experiment" conducted by the physicist Alan Sokal, who professed to being rather disappointed with the predominating intellectual quality and academic standards of postmodernist writing in the humanities, liberal arts, and cultural studies (Sokal 1996). To objectify his impression, Sokal launched an "experiment with the scholars of cultural studies." By setting up such a real-world experiment in which he made the editors and reviewers of a highly reputed journal of cultural studies his research object, he fuelled what became known as the "science wars" or "wars between the scientific cultures."[9] His experiment—or "hoax," as Sokal later called it—centred on a paper designed to appeal to postmodernists and authored by Sokal himself with the bombastic title "Transgressing the Boundaries: Towards a Hermeneutics of Quantum Gravity" (Sokal 1996). After successfully passing the review process, the manuscript was accepted by the very respectable humanities and cultural studies journal *Social Text*. After its publication, Sokal revealed that his paper was nothing but "fashionable nonsense" (Sokal and Bricmont 1998). The experiment, Sokal claimed, proved the decay of epistemic standards in the cultural studies and humanities. Sokal accused scholars in the humanities of neglecting to achieve truth, objectivity, and scientific quality. Sokal's own interpretation of his experiment did not only serve to fuel prejudices between the two cultures. On a much deeper level, it laid bare the friction between two philosophical viewpoints—realism and empiricism—which typically are associated with the natural sciences and engineering on one side of the debate and social constructivism and idealism, which are considered the canonical position of cultural scientists and scholars in the humanities, on the other side.[10] From the perspective of the *Philosophy of Interdisciplinarity*, one could conclude that the rigidness of the two thought traditions presents an obstacle to any interdisciplinary endeavour across both of them. The feasibility of interdisciplinarity cannot be taken for granted—as the notion's popularity might indicate. In sum, Sokal's experiment and the science war show how profoundly underlying philosophic positions are involved in the discourse and praxis of interdisciplinarity.

Let me next sketch another philosophical approach that envisions positive opportunities. Michel Serres (1992) seeks to renew philosophy as an academic discipline, albeit from a critical perspective. According to Serres, philosophy needs to address "interdisciplinary circulations" in the "web and knots of the sciences" and of "knowledge production."[11] In the process, a novel kind of philosophy, namely "philosophy of transport," could be conceptualized. Core elements of this interdisciplinarily oriented philosophy are "translations," "traductions," "transformations," "fluctuations," and "circulations" of knowledge, objects, and methods. For Serres, these unspecified and somewhat fluid keywords are central to characterizing "interdisciplinarity" as a form of knowledge production beyond disciplinary poles. A renewed philosophy will be engaged in the world:

> Philosophy does not just speak about the sciences, [...] it does not remain silent to the world that is based on sciences: Philosophy intervenes in the societal web of circulations. [...] Methods, models, propositions are circulating in the network; they are imported and exported, from everywhere to everywhere.[12]
>
> (Serres 1992, 8)

Every analysis of interdisciplinarity is an action, and any (act of) reflection simultaneously includes the potential revision of what is given. In line with Serres's thinking, a *Philosophy of Interdisciplinarity* can be regarded as a political endeavour since it encompasses a politics of translation, circulation, construction, differentiation, and integration of the "flows of knowledge." This broader orientation of philosophic inquiry goes beyond the 20th-century tradition of the philosophy of science.[13]

In light of the rich tradition of the philosophy of science—of which only some examples have been presented here—the reluctance of philosophers to address issues of interdisciplinarity seems incomprehensible. However, their failure to engage in the discourse on interdisciplinarity is perhaps due to the perception of the strong normative momentum of interdisciplinarity that is intertwined with non-epistemic values and, furthermore, with politics and society at large.

Motives and values

The tradition of philosophy presented above provides a first impression of the plurality of thinking in the field with which the *Philosophy of Interdisciplinarity* is concerned. Moving from the historical perspective to a more systematic approach, we are faced with another kind of plurality: the plurality of motives, values, or underlying goals. The tasks of the *Philosophy of Interdisciplinarity* are to identify and disentangle these and to render them open to critique.

Interdisciplinarians take the deficits of the disciplines and the isolation of disciplinary silos as their argumentative point of departure. Whenever "interdisciplinarity" is involved, so too are motives: Interdisciplinarians pursue—explicitly or implicitly—goals. They intend to change, renew, and restructure the sciences, the research system, and the academy or even society. Jantsch (1970), for one, as pointed out earlier, advocates a "self-renewal of the academy" and of the university structure, which he sees as the driving force for a necessary transformation of society at large. His revolutionary attitude concerning the betterment of society and the democratization of science, which was born in a time of student unrest, has met with strong opposition. Today, in line with the view held by the economist Jan Fagerberg (2005, 8), interdisciplinarity is frequently seen as—and reduced to—a resource of innovation involving the development of new technologies and long-term economic growth.[14] Such a view perfectly represents the dominant instrumental account of interdisciplinarity. Although these goals are very much present, they often remain hidden. The disregard of goals is part of the normalizing and mainstreaming process that has robbed interdisciplinarity of its critical momentum. Interdisciplinarians, nonetheless, cannot escape the normative.[15]

The notion of interdisciplinarity turns out to be a double-edged sword: On the one hand, interdisciplinarity can serve as a point of access and key catalyst for recognizing and reflecting on goals and motives of science and research in society. On the other hand, it can conceal goals and make such a debate impossible.[16] This ambivalence, or dialectic, needs to be considered and reflected upon by the *Philosophy of Interdisciplinarity*. Nevertheless, interdisciplinarity has the potential to spark deeper reflection on science and research in society. Putting this potential into practice is the guiding idea of the *Philosophy of Interdisciplinarity*.[17] A very first step in such a direction involves analysing the motives pursued by interdisciplinarians. In a nutshell, we can distinguish epistemic, economic, ethical-societal, and personal motives.[18] The respective values can be associated with truth (understanding, knowledge, insight, and objectivity—mostly curiosity-driven); utility (innovation, economic growth, and income); human and nature's well-being (basic needs, humanity, justice, democracy, peace, good life, benevolence, and sustainability); and sense-making (self-understanding, meaning, and world interpretation).

First, the *epistemic motive* frames science—and humanities—from the intra-academic perspective: Science is guided by the value of truth; it is curiosity-driven. The underlying diagnosis of the need for interdisciplinarity draws on the historically successful, functional differentiation within the academy that today reveals limits: Disciplinary boundaries turn out to hinder further advancement. Interdisciplinarity—loosely interpreted as boundary crossing and cross-fertilization—seems to be the only way to regain and ensure progress, restore knowledge production, and enable universal insight into the natural or social world. Traditionally, truth—according to Hegel's thinking—was associated with the whole and not primarily with the specialized, splintered knowledge of the disciplines. Interdisciplinarity is seen as a means to integrate and to synthesize the patchwork of disciplinary knowledge. The epistemic motive concerns interdisciplinary theories, methods, and objects in the overall architecture of the sciences.

Second, the *economic motive* does not focus primarily on the academy or on science from an intra-academic perspective: Utility is the base value by which scientific activity is framed and judged. Science is regarded as a means for obtaining and securing economic growth, prosperity, and wealth. Both Adam Smith and Karl Marx concurred, though from somewhat different angles, with Francis Bacon's viewpoint: Science is research that enables innovation and technological development; it secures international competitiveness. Accordingly, it appears to be the outstandingly powerful fundament and source of economic progress and wealth. When it comes to disciplinarity in the sciences and universities, serious deficits are manifest. The historically evolved, functional differentiation into academic disciplines does not lend itself to resolving real-world economic challenges; the utility of disciplinary knowledge is very limited. Economic practices and applications are regarded as being themselves in a certain sense inter- or transdisciplinary. In general, interdisciplinarity is seen as an instrument to overcome the disciplinary shortcomings. Considering this cluster of motives, Peter Weingart (2000, 39) speaks of "strategic" or "opportunistic interdisciplinarity."

Third, the *ethical-societal motive* is somewhat similar to the economic one, although it upholds different values. According to the ethical-societal viewpoint, research fulfils obligations within and for human and societal life. But in contrast to economic utility and technological innovativeness, the values associated with research activity are more comprehensive: They centre on the well-being of mankind, nature, and society and on sustainable development and intra- and intergenerational justice. The problems addressed in interdisciplinary research projects are therefore mainly socio-ecological, caused by the massive use of technology in society, as Erich Jantsch (1972) and Bryan Norton (2005) claim. Other scholars, such as Diana Hummel et al. (2017), additionally underscore the problematic driving forces of global capitalism. All share the view that disciplinary approaches are not adequate instruments to cope with real-world ethical-societal problems which are too complex, too wicked, and too hybrid. Interdisciplinarity is needed to tackle these problems; both normative and descriptive types of knowledge are required by political decision-makers and the public alike. These different types of knowledge have to be acquired and integrated in order to enable a sensitive, process-oriented approach to the management of complex systems.[19] Joint problem solving among science, technology, and society seems possible.

The *fourth motive* is driven by personal, metaphysical, or religious factors. The plurality of disciplinary patchworks and domain-restricted knowledge fragments creates incompatible cognitive worlds. Living in different, parallel worlds might, for some, have an almost schizophrenic impact. Interdisciplinarity is deemed a way to integrate pieces of disciplinary knowledge and connect them to a consistent or holistic picture of the entire world. The value associated with interdisciplinarity is one that is sense-making; in particular, it provides self-understanding. Such a view of interdisciplinarity shares some lines of thought with the metaphysical tradition of natural philosophy—and it also reflects some ideals of the Judeo-Christian tradition.

The foregoing list of motives is not exhaustive, but it reveals the principal grounds for interdisciplinary engagement. Moreover, it can contribute to an explicit discourse on the values associated with interdisciplinarity: The epistemic motive is guided by the value of truth; non-epistemic motives are dominated by economic values such as utility or by ethical-societal values such as human and nature's well-being, sustainability, justice, and the like; sense-making is central to personal values. In addition, the fact that multiple motives exist indicates a first plurality in our effort to clarify interdisciplinarity.

Boundaries

A philosophical approach to interdisciplinarity, as proposed in this book, naturally offers a more profound analysis than a straightforward classification of motives and values. Reflecting only on motives could easily lead to a mere descriptive approach entailing a limited view of interdisciplinarity. The *Philosophy of Interdisciplinarity* aims to critique, complement, and widen the view. One of its central objectives is to reveal underlying philosophical assumptions and fundamental convictions regarding the notion of "interdisciplinarity" —and on this basis it advances a critical perspective that opens up avenues towards sustainable knowledge within the academy (cp. Frodeman 2014).

To start with, interdisciplinarity is based, in one way or another, on disciplinarity; the term itself appears, initially, to be a derivative of disciplinarity.[20] Although the latter is not much simpler to define than the former, shifting the focus onto disciplinarity can prompt a fresh way of thinking about the field of interdisciplinarity: Interdisciplinarity urges us to rethink disciplinarity, particularly with regard to the institutional constitution of the academy, to the authority and power of disciplinary gatekeepers, and to the criteria for what counts as scientific knowledge.

In fact, the philosophy of science has shown that disciplines cannot be adequately grasped as coherent structures rooted in given domains, distinguished methods, or certain theoretical entities. Although these aspects may certainly play a role, disciplines should be perceived as historically conditioned institutional structures that are, to a greater or lesser extent, constituted by the social sphere—which is itself influenced by societal trends, historical contexts, economic interests, political decisions, and power and authority games.[21] Bearing that in mind, we can consider the two assumptions that constitute the core components of interdisciplinarity. *First*, whenever one speaks of interdisciplinarity, a *boundary premise* is present. Boundaries—which are central to any kind of differentiation, demarcation, separation, segregation, or fragmentation—are perceived to exist between disciplines as well as between academia and society. Not only do boundaries contribute to delineating disciplines, they also represent barriers and obstacles to knowledge production; boundaries are synonyms for limits and limitations.[22] *Second*, the *transcendence or transgression premise* assumes that options to overcome those boundaries are available. Interdisciplinarity aims to facilitate the transfer, circulation, synthesis, integration, or unification of disciplinary perspectives; it is typically linked to the ideal of bridging disciplines and integrating the splintered fragments of disciplinary knowledge.[23]

Taking these two complementary premises of "interdisciplinarity" results in what could be called a *boundary paradox*: the conservation *and* elimination of boundaries at the same time. The elimination of disciplinary boundaries would naturally render conservation impossible—and interdisciplinarity would dissolve. In fact, the elimination of interdisciplinarity pursuant to the elimination of boundaries has been and still is a frequently occurring phenomenon in the academic system. The historical institutionalization of computer science and informatics during the 1960s and early 1970s provides a prominent example of the gravitational pull of the normalization or mainstreaming process by which interdisciplinarity is eliminated by the creation of a new discipline. A newly institutionalized "interdiscipline" rapidly turns into a new discipline with a new regional ontology. This disciplinary pull is always a threat to interdisciplinary efforts.

Descriptively, the elimination of interdisciplinarity through the formation of a new discipline might be of interest to research fields such as the science studies or the history of sciences. From a philosophical perspective, however, we must underscore that boundaries are indispensable, and also constitutive, for any research activity labelled "interdisciplinary." Instead of "boundary paradox," a more appropriate term to describe the tension between conservation *and* elimination might be "boundary dialectic." Locating the notion of interdisciplinarity within dialectic thinking—which comes close to Hegel's *Aufhebung*—highlights that interdisciplinarity always holds a hidden critical potential. Interdisciplinarity is both dependent on disciplinarity *and* a challenge to disciplinarity. A critical-reflexive approach, as set forth throughout this book, is intrinsically bound to the recognition of and reflection on boundaries. Hence, any concept of interdisciplinarity requires a reference to boundaries and, more specifically, it requires a boundary-based dialectic concept of (a) separation and differentiation and of (b) transcendence and integration.

Employing dialectic thinking enables us to reject prominent interpretations that associate interdisciplinarity solely with integration, synthesis, fusion, unification, or holistic thinking.[24] Those positions—that also advocate overarching methods, integrative techniques, and step-by-step procedures—appear one-sided or even self-contradictory. That can be said, for instance, of the suggestion of establishing a "discipline of interdisciplinarity" and "disciplining interdisciplinarity" (Bammer 2013).

Interestingly, boundaries are an old and ongoing philosophical topic, which touches on fundamental questions about the structure of the world, the possibility of scientific knowledge, and ways to acquire that knowledge. Well-known philosophic positions embrace monistic or dualistic concepts (ontologies and epistemologies) interlaced with topics such as (non-)reductionism. Over the last thirty years, philosophers and social scientists have inquired extensively into boundaries[25] but with only very occasional reference to interdisciplinarity. Interdisciplinarians themselves, on the other hand, rarely consider boundaries and borders explicitly, although these notions are broadly taken for granted. The large overlap of the two fields carries huge potential for cross-fertilization and mutual learning. The discourse on interdisciplinarity could, for instance, undoubtedly derive benefit from the line of thought of Ulrich Beck and Christoph Lau (2004), who seek to establish a "boundary politics in the age of boundary dissolution and border elimination."[26]

To sum up, the key point is that reflection on boundaries—that is, the recognition, setting, and maintaining as well as the transcendence and transgression of boundaries; in short: *boundary work* (Gieryn 1983)—can be considered central to reflection on interdisciplinarity. Since any adequate definition of "interdisciplinarity" refers semantically to boundaries, the *Philosophy of Interdisciplinarity* needs to explicitly address boundaries and provide a conceptual framework encompassing both (a) separation or differentiation and (b) transgression, transcendence, or integration. Hence, any concept or theory of interdisciplinarity has to fulfil this twofold dialectic requirement, namely to provide a concept of separation *and* of integration.

Distinguishing different types

As set forth above, boundaries are essential. They represent central elements of the *Philosophy of Interdisciplinarity* and specifically are constitutive for advancing a critical-reflexive account of interdisciplinarity.[27] In the following, I develop a framework of different dimensions or non-disjunctive types of interdisciplinarity[28] which refers to boundaries and fulfils the two related requirements.

While the (*intensional*) semantic core of "interdisciplinarity" consists, on a general level, of boundaries, the different types will show that interdisciplinarity is a multifaceted phenomenon. That is to say, from a philosophic viewpoint, the *extension* or scope of the term has to be characterized by a plurality: a plurality of types united in a semantic core of boundaries. Such a framework—being central to the *Philosophy of Interdisciplinarity*—is not an end in itself but will be used in order to analyse, assess, and critique interdisciplinary research programs. By employing distinctions that are well established in the tradition of philosophy—such as those between objects (ontology), knowledge/theory (epistemology), methods (methodology), plus one central additional aspect, namely problems—we can identify four types or dimensions of interdisciplinarity, and we can relate this result to what has become known as transdisciplinarity.[29]

First, an object-oriented or ontological type of interdisciplinarity can be defined in terms of objects, entities, or structures of reality such as the human brain, the evolution of the earth, the ozone hole, nanoparticles, nuclear power plants, personal computers, the internet, skyscrapers, water supply systems, or military infrastructures. The basic assumption with regard to this kind of interdisciplinarity is that the historical, functional differentiation of the academy into institutionalized disciplines does not seem absolutely contingent. Rather, the differentiation mirrors aspects of the nature of the things themselves. Interdisciplinary objects are deemed to be located within or built into the deep structure of reality. Edmund Husserl, Nicolai Hartmann, Alfred North Whitehead, and others argue, for instance, in support of a structurally layered concept of reality according to which interdisciplinary objects would lie on the boundaries between different micro-, meso-, macro-, and other cosms or within the border zones between disciplines. Some examples include brain-mind objects, nano-objects, or entities of synthetic biology. To advocate this position, one must presuppose an ontological realism or at least a real-constructivism[30] concerning objects, interlaced with a layered concept of reality, and, based on this, an ontological non-reductionism.[31] Old and ongoing issues about ontological monism, dualism, and pluralism emerge in this debate. Interdisciplinarity according to this view is not concerned mainly with knowledge, methods, or research goals but above all with a reality that is assumed to be independent of humans. A minimal realist view of the things is involved. More recent versions of this position do not assume the timeless, somewhat Platonic existence of objects.[32] New interdisciplinary objects are constructed and created through the massive spread of technologies or are cognitively constructed by the sciences themselves. Examples include the hole in the ozone layer, high-frequency trading on the stock exchange, or virtual objects in computer science. The massive spread of human-constructed objects, sometimes labelled "human-created nature," also supports the observation that we are witnessing an epochal break with regard to ontology, as Martin Carrier (2011, 51) argues.

Second, a knowledge and theory-oriented type of interdisciplinarity sees epistemological aspects as the central criterion. The focus lies on knowledge, propositions, theories, models, and concepts[33] and not primarily on objects or methods. The crucial questions in this case are the following: How can we demarcate interdisciplinary knowledge from disciplinary knowledge or from non-scientific knowledge? How should we specify interdisciplinary theories, models, laws, descriptions, and explanations? Do these provide a specific conceptual understanding of the objects under consideration? Potential candidates for interdisciplinary theories or concepts include meta-theories, which can be applied to describe very different disciplinary

objects. According to this understanding, an interdisciplinary theory highlights structural similarities between the properties of different objects from
various disciplines. Systems theory is one of the most prominent examples
of an interdisciplinary theory, as is cybernetics and, to a certain extent, some
variants of the theory of evolution. Furthermore, strong arguments have
been put forward over the last fifty years claiming that, on a deeper level,
we are witnessing the emergence of novel interdisciplinary theories, specifically self-organization theories, which provide an evolutionary, dynamic,
self-organizing understanding of the entire world. As elaborated above, the
basic requirement is that an interdisciplinary theory must not be reducible
to a disciplinary one; that is, interdisciplinary theories do not fit into disciplinary frameworks. An epistemological non-reductionism of interdisciplinarity with regard to disciplinary theories is the most compelling stance.
Theory-oriented interdisciplinarity questions the ideal of the covering-law
model and the feasibility of grand unification based on the subsumption of
all phenomena under a disciplinary law. Given the historical development
of the sciences and, in particular, of such concepts as systems theory, it is
overall evident that theory-oriented interdisciplinarity does not constitute
an epochal break or a rupture in the theoretical core of the sciences.

Third, interdisciplinarity is often viewed from a methodological angle
and frequently regarded as a challenge to scientific methodology.[34] A
method-oriented type of interdisciplinarity can be identified. Methodology generally refers to knowledge production, the research process, rule-
based actions of scientists, procedures of inquiry, and the languages used
therein. In methodology, the main issue is how, and by which rules and
procedures, can we obtain knowledge and insight. This procedural understanding of the sciences—science as research—is sometimes called context
of discovery or the research form of science to distinguish it from the context of justification, which refers to knowledge, theories, or propositions.[35]
Rough, classical categorizations distinguish between empirical and hermeneutic, nomothetic, and ideographic methods as well as (more generally)
between the methods of the natural sciences and those of the humanities
or between explanation and understanding. With respect to interdisciplinarity, some of the central questions are the following: Do interdisciplinary
methods and actions exist? Is there a specific context of discovery prevalent
in interdisciplinary projects? Which validity, evidence, and quality criteria
can be applied to the results of interdisciplinary projects? Do they differ
from those in disciplinary projects? Interdisciplinary methods are thought
to be irreducible to disciplinary ones. Outstanding prospects are ascribed
to those interdisciplinary methods that facilitate the transfer of knowledge
between disciplines and also to those that combine descriptive, normative,
and abductive methods of reasoning beyond disciplinary havens. In addition, other scholars see interdisciplinarity as a (transdisciplinary) method to
bridge the gap between the academy and society—in other words, methods
are interdisciplinary when they enable, facilitate, and foster knowledge production between the academy and society in one way or another.

Fourth, we need to add another level of reflection—since it is quite common to speak of interdisciplinarity in connection with addressing certain problems that are deemed beyond the scope of a specific discipline and even outside the academy. This type of interdisciplinarity is often described as problem-oriented, purpose-driven, or issue-focused or—with a slightly different meaning—as transdisciplinarity (see also Figure 2.1); it is important to point out that transdisciplinarity should not be restricted to cases where stakeholders or lay people are involved in the process of knowledge production.

Compared with the three other types of interdisciplinarity, the fourth approach frames science and research from a more comprehensive perspective. It centres on problems and issues, and it includes the goals, purposes, initial conditions, and research agendas of scientific activities. This approach concurs, for instance, with the thinking of Jürgen Habermas (1971), who adverts to the guiding interests of scientists and their research agendas. It is typically based on the assumption of a teleological structure governing the process of knowledge production: A trajectory is presupposed to exist from the point of agenda setting—where the problems are perceived or defined—to the anticipated results.

The problem dimension, or the "context of problems," notably the will to know, precedes both the "context of discovery" and the "context of justification" (i.e., the methods and theories). Despite the obvious significance of the points of departure—in the field of interdisciplinary research and also in disciplinary research—philosophers have surprisingly rarely acknowledged such a broader and more appropriate view of the sciences. Their reluctance might stem from a fear that problems, because of their being obviously value-laden, cannot be separated from the social dimensions of research activity. Outside the mainstream of philosophy of science, thought-provoking approaches have been pursued under the label of the new field of social epistemology (Fuller 2002)—although scholars in this field have not inquired in-depth into problems and agenda-setting procedures. The same holds for innovative fields of the social sciences in which scholars have addressed so-called "wicked problems" (Rittel and Webber 1973). The structure of problems, however, has hardly been clarified and understood: Problems therefore remain a "no man's land" in terms of explicit reflection (see Chapter 5).

In regard to interdisciplinary problems, it could at first sight (a) be generally assumed that they may be merely epistemic in nature; they can emerge through intra-academic progress and require an interdisciplinary effort within the sciences (see Figure 2.1). For instance, problems in the field of physical cosmology demand collaboration among physics, chemistry, geology, and computer science. Solving these problems is of interest to the sciences and serves their truth seeking but does not have wide-ranging relevance for society at large. In addition, (b) interdisciplinary problems can emerge within the economic or business field. Finding solutions to these trans-scientific or extra-scientific problems is guided by the value of

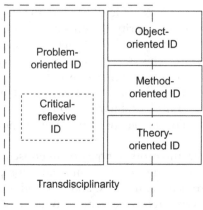

Figure 2.1 Landscape of the definitions employed in this book. We can distinguish four types or dimensions of interdisciplinarity (ID): object-oriented ID, method-oriented ID, theory-oriented ID, and problem-oriented ID. The last focuses on societally or ethically relevant (extra-scientific) problems and can be regarded as instrumentalist. Some problem-oriented interdisciplinary projects are also critically reflexive: The latter category is a subset of problem-oriented ID. In addition to a mere means-centred instrumentalist approach, critical-reflexive (problem-oriented) ID also involves reflection on and, if deemed necessary, the revision of the problems, goals, purposes, or values of research agendas. Of most interest is the overall relation of ID to transdisciplinarity (TD) insofar as TD is a very popular notion. The latter should not be restricted to or defined by referring just to the involvement of (extra-scientific) lay people or stakeholders (see this Chapter 2). Since TD addresses trans-epistemic, extra-scientific, or real-world issues (= mode 2, trans-science, post-normal science, technoscience, and the like), TD and ID (in particular: TD and problem-oriented ID) need to be distinguished. Some disciplinary issues are transdisciplinary but *not* interdisciplinary and, more specifically, *not* problem-oriented interdisciplinary (e.g., the design and construction of a bridge). In fact, many engineering challenges are to be considered disciplinary *and* transdisciplinary. Conversely, certain interdisciplinary objects (object-oriented ID) cannot be considered transdisciplinary, such as the cosmos (in particular, the cosmic evolution of the universe) or the human brain and its (self-)consciousness. Since problem-oriented ID always refers to real-world or trans-scientific, societally/ethically relevant problems, problem-oriented ID is effectively a subset of transdisciplinarity. In addition, some interdisciplinary objects cannot be classified as problem-oriented, such as (techno-)objects on the nano-scale that extend across the borders between physics, chemistry, biology, informatics, material engineering, and others. Nano-scale projects are very technically centred and therefore cannot be regarded as being problem-oriented, although they are clearly transdisciplinary since they are associated with extra-scientific or trans-epistemic goals. Therefore, the notion of problem-oriented transdisciplinarity is one that makes sense. The point to recall is that projects may be disciplinary *and* transdisciplinary (e.g., bridge construction) or interdisciplinary *and* non-transdisciplinary (e.g., the cosmic evolution). Transdisciplinarity is not the counterpart to (and not a disjunctive form of) disciplinarity.

economic utility and is often motivated by the growth imperative of share-holder value.

However, although both kinds of interdisciplinary problems clearly pose challenges to interdisciplinary collaboration, they are usually *not* what interdisciplinarians have in mind when they speak of "problem-oriented," "issue-initiated," or "purpose-driven" interdisciplinarity. Let us therefore discuss the concept of problem-oriented interdisciplinarity and distinguish the notion of problem used in this context from the two kinds of interdisciplinary problems outlined above. Of most interest is the overall relation of interdisciplinarity to transdisciplinarity in general.

Problem-oriented interdisciplinarity is always transdisciplinary (ad a): Although the advocates of problem-oriented interdisciplinarity do not deny the existence and scientific relevance of *intra-epistemic* problems,[36] they stress the transdisciplinary nature of the type of problems they focus on and refer to interdisciplinarity as transdisciplinarity.[37] In general, the notion of transdisciplinarity underscores a trans-academic or extra-epistemic orientation of knowledge production: Science produces knowledge *within* and *for* society or the economy.

Now, when interdisciplinarians describe their approach as transdisciplinary, they are not denying that traditional disciplines such as certain engineering or technical sciences can be seen as being transdisciplinary-oriented.[38] The key point made by interdisciplinarians in labelling their work "transdisciplinary" is that the nature of the problems addressed in their projects is overall trans-academic: These problems—sometimes referred to as "real-world problems"—are not deemed relevant only by the peers of a certain scientific discipline (e.g., engineering scientists) or by scholars of the academy but by society at large.

Furthermore, *problem-oriented interdisciplinarity is more specific than transdisciplinarity; the former is a subset of transdisciplinarity* (ad b). With regard to the second kind of problems discussed above—those in the economic or business realm—it is to be noted that, while economic problems can obviously have a transdisciplinary dimension and often require interdisciplinary research, the notion of problem-oriented interdisciplinarity is generally linked to societal and ethical motives. These problems are primarily ones that represent challenges to society at large, such as the climate change, the loss of biodiversity, the limits of energy resources, the new regional wars, the threat posed by atomic weapons, the nuclear waste of power plants, the threat to human health by environmental pollution, and the global injustice surrounding the worldwide distribution of wealth. For instance, all issues that hinder us from pursuing the goals of sustainable development represent the kind of problems addressed by problem-oriented interdisciplinarity. The notion of transdisciplinarity is therefore much broader in scope and more unspecific than problem-oriented interdisciplinarity.[39] Furthermore, transdisciplinarity should not be reduced to stakeholder and lay people involvement, as Jaeger and Scheringer (2018) convincingly argue.

To provide a deeper and more detailed explication, I will now further distinguish between two kinds or modes of problem-oriented interdisciplinarity. One mode involves reflection on and, if deemed necessary, the revision of the problems and goals of research agendas—which includes an argumentative justification of the relevance of the problems addressed. Contrary to this first mode, the second one accepts problems as being simply given. Throughout this book, the former will be termed the *critical-reflexive kind* of problem-oriented interdisciplinarity whereas the latter is primarily an *instrumentalist* or *strategic account*. The relation between the two modes will be elaborated on; it will be shown that the critical-reflexive approach can be seen as a subset of the instrumentalist account of the fourth type of interdisciplinarity, namely of problem-oriented interdisciplinarity. The other three types of interdisciplinarity discussed above—the object-, theory-, and method-oriented type—share an instrumental dimension with the instrumental mode of problem-oriented interdisciplinarity.

Taking stock of the arguments presented so far, "interdisciplinarity" is semantically justifiable if, and only if, at least one of the four types or dimensions can be ascribed to it. At the same time, the four types are not exclusive or disjunctive. It is possible that a specific research project fulfils more than one dimension: For example, it may be both problem- *and* object-oriented. In a later subsection of this chapter, we will see how different philosophic convictions determine how the different types of interdisciplinarity are seen to be related.

Examples

The framework of the four types of interdisciplinarity can be further illustrated by some popular examples of research programs that are considered "interdisciplinary." These examples also give further substance to the classification advanced by the *Philosophy of Interdisciplinarity*.

First, let us consider interdisciplinary objects and the ontological type of interdisciplinarity. Nanoresearch is one of the most prominent fields that claim to be interdisciplinary—for instance, in an influential report presented by the US National Science Foundation (Roco and Bainbridge 2002).[40] In 1959, the physicist Richard Feynman stressed the presence of "white and unconquered domains" on the "disciplinary map of sciences": There seems to be "plenty of room at the bottom" (Feynman 2003).[41] According to Feynman, nano-objects are located between the microscale of quantum physics and the mesoscale of chemistry and biology. Some of them are designed, constructed, and created by researchers; others existed or came into existence independently before the emergence of nanoresearch but have now been discovered or brought under control. Nano-objects are interdisciplinary in nature insofar as they lie on (or between) the boundaries of scientific disciplines, whereas the boundaries themselves are thought to mirror the deep ontological structure of reality.[42] Interdisciplinary nano-objects seem to be the unifying core and umbrella notion encompassing the heterogeneous fields of nanoresearch and nanotechnology, which include electron-beam and ion-beam fabrication, molecular-beam epitaxy, nano-imprint lithography, projection electron microscopy, atom-by-atom manipulation, quantum-effect electronics, semiconductor technology, spintronics, and micro-electromechanical systems.[43] In these examples, interdisciplinary objects are an essential part of reality on an ontological level: The nano-objects, constructed in a joint effort by the disciplinary fields of physics, chemistry, biology, and engineering sciences, are today regarded as technoscientific objects. Nano-objects—and object-oriented interdisciplinary research—have not yet been perceived by philosophers and social scientists, with the exception of Martin Carrier (2011), as a type of interdisciplinary engagement.

Similarly, the objects with which neuroscience is concerned are located on various boundaries between the disciplines (i.e., between the natural sciences, social sciences, and the humanities). Interdisciplinary neuroscientific objects are much more complex than nanotechnological objects. Furthermore, socio-technical objects, such as the water supply system, the internet, or the hole in the ozone layer, are further examples of interdisciplinary objects that are representative of object-oriented interdisciplinarity.

Second, let us now turn our attention to epistemological interdisciplinarity, which relates to knowledge, theories, models, and concepts. Systems theory and complex systems are examples of interdisciplinary theories and interdisciplinary knowledge. Cognitive integration and a theoretical synthesis of knowledge—which avoids the trap of reductionism—are goals that have been partly attained in this area. Other interdisciplinary concepts, which are very similar to and interlaced with complex systems theory, include self-organization theories, dissipative structures, synergetics, chaos theory, nonlinear dynamics, fractal geometry, and catastrophe theory (Mainzer 1996; Schmidt 2008a, 2011a, 2019). Most of these concepts were established in the late 1960s and early 1970s, although some foundational work dates back to the late 19th century. Hermann Haken (1980), for instance, regards his research field of synergetics as an "interdisciplinary theory of general interactions." Erich Jantsch (1980) views self-organization theories and the general concept of evolution as a "unifying approach" with multiple "implications to the sciences and the humanities." Edward O. Wilson (1998) anticipates a new "consilience" entailing a "unity of knowledge," for instance, through research programs in socio-biology. Klaus Mainzer (2005, v) identifies within complex systems theory "the basic principles of a common systems science in the 21st century, overcoming traditional boundaries between natural, cognitive, and social sciences, mathematics, humanities and philosophy." On closer scrutiny, however, this type of theory-oriented interdisciplinarity, which could also be characterized as meta-disciplinary or non-disciplinary,[44] is not as novel as it might appear. It is also found in works from the 1950s. At that time, the physicist and philosopher Carl Friedrich von Weizsäcker (1974, 23) coined the term "structural sciences."[45] As Weizsäcker writes, structural sciences "study their objects regardless of disciplinary origin and in abstraction from disciplinary allocation." Weizsäcker had in mind concepts such as information theory or cybernetics.

Today, structural sciences have been extended and enriched by complex systems theory, which investigates nonlinear, unstable, and chaotic behaviour in dynamic systems and describes processes as they evolve over time, such as pattern formation, self-organization, critical behaviour, bifurcations, phase transitions, structure breaking, and catastrophes. Complex systems theory addresses old questions concerning the emergence of new phenomena and about novel properties, patterns, entities, and qualities. One important lesson provided by this interdisciplinary concept for all sciences is the fundamental role of instabilities in nature, technology, and even in social processes.[46]

Third, we will go deeper into the methodological type of interdisciplinarity. Bionics or biomimicry could be regarded as prominent examples of an interdisciplinary method.[47] These fields claim to provide a method of transfer between two disciplines: from biology to engineering sciences and probably (though this is usually not acknowledged) vice versa. The central idea of bionics is, it is maintained, to "learn from nature" in order to "inspire technological innovations" and to "optimize artifacts and technical processes" (Benyus 2002). Nature is seen as being productive. As such, it serves as a source of inspiration for inventions that can be used for the design and construction of new technical systems. The proponents of bionics are convinced that nature "reaches its goals efficiently and economically, with a minimum of available energy and resources. The experience available in nature can be applied to conduct technological research and development" (Hill 1998). Interdisciplinarity in the methodological sense is here based on a kind of "translation" or "transfer" between nature and technology—more specifically between certain framings, representations, perceptions, understandings, or models of nature and of technology. "Learning from nature" therefore means learning from *models* of nature: Nature is not simply given but is constituted or constructed, as Immanuel Kant argued. The models of nature built in the field of bionics are based on the perspective of engineering sciences. For example, a robot may be a technical model of an ant and therefore mimic the ant, but at the same time the ant is modelled by the bionics researcher from a technological perspective. Construction and reconstruction—in this case technology/engineering science, on the one hand, and biology, on the other—are, at least to some degree, merged since the goal of bionics is not only to produce knowledge but also to create technological artefacts. Bionics can be seen as a paradigm of a *technoscience* based on an interdisciplinary method of transfer across the *border or trading zone* between biology and engineering (Galison 1996; Gorman 2010).

There are further examples of interdisciplinary methodologies besides bionics. Econophysics, which methodologically organizes a knowledge transfer between physics and finance/economics, is another paradigm of the methodological type of interdisciplinarity that is very similar to bionics.[48]

Further examples encompass transdisciplinary methods employed in problem-oriented interdisciplinary projects aimed at organizing and managing the knowledge transfer and production between the extra-scientific and academic participants (Pohl and Hirsch Hadorn 2007; Bergmann et al. 2012).

Fourth, the most far-reaching type of interdisciplinarity is problem-oriented interdisciplinarity, often referred to as transdisciplinarity—although not all transdisciplinary projects are to be considered problem-oriented or even interdisciplinary (see Figure 2.1). As outlined earlier, interdisciplinary problems can in general be inner-academic ones emerging in the curiosity-driven domain of truth seeking within the sciences, such as problems regarding the origin of life on earth or dealing with the characteristics of consciousness of the human brain. However, these are not the kind of problems that scholars specifically mean when they use the expression "problem orientation." So we need to look beyond the curiosity-driven academic field and beyond an economic perspective on interdisciplinarity. Technology assessment, sustainability science, and social ecology are paradigms of problem-oriented interdisciplinary approaches. Research in these fields starts from the perception of pressing interdisciplinary problems that are seen as being societally and ethically relevant.[49] These fields aim to obtain systems, target, and transformation knowledge (Pohl and Hirsch Hadorn 2007)—accompanied by consideration of the possible societal impact of new and emerging technologies (e.g., side effects, risks, and potentials)—in order to address prospective problems as early as possible.

Problem-oriented interdisciplinarity as a specific type of interdisciplinarity has its own history. We may recall that Jantsch, at an OECD conference in the early 1970s, called for "inter- and transdisciplinarity" not only in an academic context but also for societal and ethical purposes. Jantsch accused the university and academic systems of being incapable of addressing the pressing real-world problems such as warfare (involving atomic, chemical, or biological weapons); environmental problems such as global warming and the loss of biodiversity; waste production, disposal, and contamination; shrinking natural resources; problems with water and food quality; and anthropological problems in connection with the ambivalence of biomedical progress.

This book is dedicated mostly to problem-oriented inter- or transdisciplinarity; further examples—in particular, ones seeking to obtain a different view of nature and the environment—will be discussed later on.

Schools of thought

The four types of interdisciplinarity sketched above can be regarded as *ideal* types, which, granted, do not occur exclusively in particular scientific practices or programs, and they are by no means disjunctive in the sense that a research practice or program can be subsumed under one type or another. For example, a specific research program claiming to be interdisciplinary can be both problem- and method-oriented. Nonetheless, one type of interdisciplinarity will typically dominate whereas the other types are seen as derivatives that are related in one way or another to the principal type. Moreover, which of the types of interdisciplinarity will take precedence over which is undeniably open to discussion and subject to justification. A method-oriented interdisciplinary research program, for instance, can be seen as a consequence of a certain problem orientation or vice versa.

In light of the foregoing considerations, the *Philosophy of Interdisciplinarity* endeavours to provide additional arguments to confirm the existence and prevalence of the four types of interdisciplinarity. Most interestingly, one's philosophical background conviction is what, consciously or unconsciously, predetermines which of the four types one might consider most relevant and which types one might see as inferences, derivations, or mere consequences.[50] The *Philosophy of Interdisciplinarity* enquires into the philosophical background and into the various implicit philosophies influencing the discourse on interdisciplinarity.

With their primary focus on objects, things, and artefacts, realists and (to some extent) empiricists, new experimentalists, and real-constructivists first assume the existence of given or constructed objects which can be cognitively perceived from an objective angle. In this object-centred account, interdisciplinary research—like any research activity—commences directly on the ontological level. The need for interdisciplinarity stems from the essential structure of the objects located beyond or across disciplinary boundaries. Given the very existence of interdisciplinary objects, one has to select or develop adequate (namely interdisciplinary) methods; the methods are assumed to be prescribed by the structure of the objects. Furthermore, interdisciplinary knowledge—for realists, empiricists, and others—originates in or results from the interdisciplinary objects themselves. In sum, a realist or empiricist position is central to being able to defend object-oriented interdisciplinarity. Those who deny the plausibility, soundness, or justification of the object-oriented view of interdisciplinarity are at the same time attacking some of the central assumptions of realism, empiricism, or new experimentalism.

Rationalists, *second*, tend to frame interdisciplinarity by referring primarily to knowledge, theories, models, concepts, theoretical entities, or even mathematical structures. According to such a rationalist perspective, interdisciplinarity becomes necessary because of the increasing fragmentation of knowledge and the lack of unity: Interdisciplinarity is regarded as an attempt to counteract this development. Its goal is to contribute to a broader view of the things by bridging, synthesizing, integrating, or unifying various knowledge fragments from different disciplines and subdisciplines. The guiding ideal is to provide a coherent picture of the whole of reality or at least of those things that are regarded as a central part of reality. In contrast to the belief underpinning object-oriented interdisciplinarity, the different academic disciplines are *not* assumed to mirror different natures of the objects, but are defined by a specific kind or corpus of knowledge, theories, or models. The theory-oriented type of interdisciplinarity certainly seems the most ambitious given that it requires a process of cognitive integration or even unification. According to this understanding of interdisciplinarity, theories or concepts have first priority and precede the selection of methods, the framing of objects, and the definition/constitution of problems. Object-, method-, and problem-oriented interdisciplinarity are therefore seen as derivatives of theory-oriented interdisciplinarity.

Many interdisciplinarians—notably methodological constructivists,[51] scholars from science and technology studies, and some pragmatists—*third*, tend to reflect on methods, rules of knowledge production, practical procedures, or heuristics. They regard science as a method-based action (i.e., as a research activity). Accordingly, interdisciplinarity is defined by methods: Interdisciplinary methods are considered to be non-reducible to disciplinary ones. Interdisciplinarity from this perspective challenges researchers to set up new procedural rules, to create a vocabulary overlapping disciplines, to establish novel validity or evidence criteria, and, in addition, to organize the collaboration among the disciplines, and to institutionalize interdisciplinary research processes. The recent predominance of disciplinary cultures—and, more specifically, of disciplinary orientations, validity criteria, heuristics, habits, vocabularies, and languages—poses major obstacles to interdisciplinarity. The primacy of methods means that interdisciplinary problems, objects, and knowledge are mere derivations or consequences of the respective methods.

A different approach to interdisciplinarity, *fourth*, is taken by pragmatists, utilitarians, critical theorists, and others such as political philosophers and many ethicists insofar as they refer to problems, goals, purposes, and interests. According to them, the need for interdisciplinarity is due to the emergence of pressing problems that do not fit into the disciplinary differentiation of the academy. Holders of these viewpoints evaluate interdisciplinarity by its ability to pragmatically define or address problems. Typically, the problems in question have societal and ethical relevance. The advocates of problem-oriented interdisciplinarity see the framing, construction, or reconstruction of objects as a mere consequence of the perception or constitution of problems. The methods, and also the ensuing knowledge, are regarded as derivations.

To summarize, philosophical schools of thought serve as lenses through which one views both the disciplinary and the interdisciplinary scenery. They determine what meaning and significance one is willing to attach to interdisciplinarity—and which order of priority, hierarchy, or chain of inference of the different types of interdisciplinarity one is willing to subscribe to. Given the relevance and prevalence of the four positions on interdisciplinarity, an elimination of the plurality of understandings and their reduction to a single meaning—beyond the twofold, dialectic reference to boundaries—is not feasible. The plurality of notions of interdisciplinarity mirrors the plurality of the different intellectual traditions and schools of thought in philosophy. Therefore, the debate on interdisciplinarity is, in a broader sense, philosophical in nature.

Conclusion and prospects

The typology of interdisciplinarity presented in this chapter is intended to serve as an orientation framework. With reference to established positions in philosophy, different types of interdisciplinarity can be distinguished: the object-oriented type, the theory-oriented type, the method-oriented type, and the problem-oriented type, the last of which can be regarded as an effective subset of what is known as transdisciplinarity. On a more general note, transdisciplinarity opens science to society; it is not an oxymoron to see transdisciplinary disciplines. More specifically, the problem-oriented type of interdisciplinarity, as a subset of transdisciplinarity, considers ethical and societal aspects. Thus, we can also speak of problem-oriented transdisciplinarity.

Above all, the semantic core of interdisciplinarity is tightly connected to boundaries. This entails a dialectic relation between boundary setting and preserving *and* at the same time transcending and overcoming boundaries. The acknowledgement of an underlying dialectic relation gives substance to a twofold requirement that is intertwined with both non-reductionism *and* integration (or synthesis): Interdisciplinary objects, theories, methods, and problems are deemed to be irreducible to disciplinary ones. Ontological, epistemological, and methodological boundaries, as well as the boundaries of the academic system, are seen as obstacles or barriers to various kinds of reductionism. On the other hand, boundary-crossing is an indispensable aspect of interdisciplinarity. In fact, the bridging of boundaries or the transfer of knowledge across boundaries can be viewed as a kind of integration, synthesis, or reduction.

Therefore, interdisciplinarity is inherently linked with a philosophical position that can be called *integrative non-reductionism* or *non-reductive integrationism*. It shares much with a newly proposed concept, namely *integrative pluralism* (Mitchell 2009).[52]

Notes

1 According to this viewpoint, not the objects appear to be messy and vague, but rather the terms or propositions in which we represent them.
2 See, among others, Balsinger (1999, 2005), Hubig (2001), Grunwald and Schmidt (2005), Schmidt (2008b), Frodeman (2010), Krohn (2010), Jungert et al. (2010), Wechsler and Hurst (2011), and Gethmann et al. (2015).
3 However, philosophy of science should not be viewed in the limited sense in the tradition of analytic philosophy, but from a broader perspective including social epistemology, political philosophy of science, ethics of science, philosophy of technoscience, history of science, and more.
4 See the analysis by Krohn (2010), Gethmann et al. (2015), and Krohn et al. (2017).
5 For a critique, see Holbrook's (2013) line of argument.
6 My translation (J.C.S.).
7 My translation (J.C.S.).
8 Snow's thesis was later supported by an empirical study conducted by the philosopher Schurz (1995).
9 An introduction to the "science wars" is given from various perspectives in Bammé (2004) and Segerstrale (2000).

10 The "science wars" later abated when Hacking (1999) advocated a pluralist understanding of constructivism and of realism. Hacking showed that Sokal's pessimistic conclusion is based on rough assumptions and prejudices, for example, on some elements of a naive scientific realism.

11 My translation from the German version (J.C.S.).

12 My translation from the German version (J.C.S.).

13 This is in line with Frodeman's (2010, 2014) agenda to renew (Anglo-American analytical) philosophy.

14 For a critique, see Maasen (2010).

15 This is an experience similar to that facing scientists during a revolutionary phase in which the paradigm, the disciplinary matrix, and the underlying norms dissolve.

16 See, for instance, Weingart and Stehr (2000).

17 As such, the *Philosophy of Interdisciplinarity* can be seen as a central element of a critical-reflexive interdisciplinary practice. That is to say, the critical-reflexive type of interdisciplinarity includes the *Philosophy of Interdisciplinarity*.

18 A cognate distinction that refers to "values" is presented in Gethmann et al. (2015) and Machamer and Wolters (2004). For a more specific account of values regarding sustainable development goals (SDGs), see Norton (2015).

19 See the argumentation offered in Norton (2005).

20 See also Balsinger (1999, 2005), Turner (2000), and Jungert et al. (2010). From an analytic perspective, this point is inescapable, although Krohn (2010, 33) advances a thought-provoking understanding of interdisciplinarity without referring to disciplines.

21 New fields of philosophical inquiry such as social epistemology, science and technology studies, and technoscience studies paint a very colourful and diverse picture of the various disciplines and subdisciplines.

22 In this vein, Frodeman (2014, 3) sees "the notion of limit" as "a core meaning" of inter- or transdisciplinarity.

23 An extreme kind of integration is known as reduction.

24 Arguments against the integrationist stance are provided by Holbrook (2013).

25 An explicit reflection on boundaries can be found in Star and Griesemer (1989) and Löwy (1992)—and, from a different angle, in Beck and Lau (2004).

26 My translation (J.C.S.).

27 Since acknowledging and reflecting on boundaries are not only a prerequisite for, but also an indispensable part of, any good interdisciplinary practice, the *Philosophy of Interdisciplinarity* can be seen as a central and essential component of interdisciplinary practice.

28 A thought-provoking approach towards a continuum of types of interdisciplinarity has been developed by Szostak (2015) and Szostak et al. (2017).

29 A draft of this typology is developed in Schmidt (2005) and Schmidt (2008b).

30 The position of real-constructivism (see Schmidt 2011b) is not fully developed in philosophy, although the "new experientalism" has argued in favour of it (e.g., Hacking 1983, 1999).

31 Ontological reductionism holds that the world consists of atoms or other fundamental material entities ("materialism") or, on the contrary, of mental entities ("idealism").

32 As the sciences progress and scientific institutions change over time, objects that were previously interdisciplinary can be shifted into domains of new disciplines with their novel regional ontologies.

33 Hübenthal (1991) identifies "concept interdisciplinarity" as a specific type of interdisciplinarity which is concerned with systems theory, cybernetics, synergetics, information theory, and others. For a general consideration of complexity and systems theory, see Kline (1995).

34 See, for example, Pohl and Hirsch Hadorn (2007), Hirsch Hadorn et al. (2008), or Bergmann et al. (2012).

35 This highly disputed distinction can be traced back to the Vienna Circle and Hans Reichenbach.

36 A distinction between *intra*-scientific and *extra*-scientific (transdisciplinary) problems is usually presupposed. It can be traced back to heated debates in the philosophy of science on the subject of internalism and externalism (cf. Böhme et al. 1983). Transdisciplinary problems are considered to be science-*external*. A split is assumed between science and society. For a discussion of this issue from a sociological perspective, see Cozzens and Gieryn (1990).

37 This view is in line with the thinking of Gibbons et al. (1994).

38 It would be absurd, indeed, to conceive of engineering research on renewable power plants, electric cars, fuel cells, railway bridges, lighter materials, or atomic weapons as a mere epistemic or inner-academic enterprise.

39 As outlined, "problem-oriented transdisciplinarity" is a subset of "transdisciplinarity." Therefore, we can also talk about "problem-oriented transdisciplinarity." The two terms—"problem-oriented interdisciplinarity" and "problem-oriented transdisciplinarity"—are synonyms.

40 For a more detailed analysis, see Chapter 3.

41 Nanoresearch is based on technological advancements: scanning tunnelling microscopy and the atomic force microscope, which stem from developments in the early 1980s.

42 Indeed, there may be an ontological boundary or a boundary zone between the microscale and the mesoscale in which the given or constructed objects can be located.

43 See also Chapter 3. Since many problems in nanoresearch are driven by extra- or trans-scientific/-epistemic motives or needs—that are typical for engineering sciences—most areas of nanoresearch are also to be characterized as transdisciplinary.

44 Other authors also refer to the framework of systems theory when considering a conceptual foundation of interdisciplinarity (Kline 1995).

45 Original German term: *Strukturwissenschaften*; see, for example, Küppers (2000, 89ff.).

46 See also Chapter 7 in this book.

47 For an introduction to this field, see Benyus (2002), Nachtigall (1994), Maier and Zoglauer (1994), and Schmidt (2002). Klein (2000, 3f.) speaks of "borrowing" with regard to methods.

48 For an introduction to this field, see Mantegna and Stanley (2000) and McCauley (2004).

49 See, for instance, Gethmann (1999), Decker (2001, 2004), Chubin et al. (1986), and Lingner (2015).

50 Even interdisciplinarians who never explicitly engage with philosophical thought traditions have their implicit philosophies on which their perceptions, actions, and judgments of interdisciplinarity are based.

51 This is a philosophic position developed in German-speaking countries (Lorenzen 1974; Janich 1984; Janich 1992) that has not been broadly recognized by the international community of philosophers of science.

52 Admittedly, integrative non-reductionism would require further elaboration—which, alas, goes beyond the scope of this book (aspects are discussed in Schmidt 2015a).

3 Politics and research programs
Addressing the knowledge politics of interdisciplinarity

Knowledge politics

Interdisciplinarity—including its central cognate: transdisciplinarity—is, above all, a political term:[1] It is at the centre of recent knowledge politics and political epistemology.[2] It is not a term signifying simply a methodological or management challenge to the sciences. Whenever the notion of "interdisciplinarity" shows up in public as well as in scientific discourses, we are debating—explicitly or implicitly—the future of scientific knowledge and also the role and function of academic institutions and the university in society. The *Philosophy of Interdisciplinarity* is thus concerned with politics: What do our late-modern societies need to know—and for what purposes? What fundamental insight into nature is desirable and therefore worth promoting? What kind of knowledge is necessary for the future of our societies? What kind is of minor relevance?[3] What claims of truth and expertise are made (or should be made) and how do they impact politics and decision making?

Although science (research) and technology (development) politics has been a familiar theme since the time of Francis Bacon and, in particular, predominated the "big science" era, as exemplified by the Manhattan project (Bush 1945; de Solla Price 1963), *knowledge politics* is a relatively new field of intensive political activity that has rapidly emerged over the last 50 years. It is one that urges us to question the traditional conception of curiosity-driven knowledge production and value-freeness of scientific facts, which for a long time dominated the philosophy of science:[4] Knowledge is always interlaced with human interests—in one way or another.

From the 20th century on, we have been experiencing a politicization of knowledge. Moreover, in today's neoliberal times and with the emergence of the *knowledge economy*, knowledge is regarded as a commodity, resource, and means of achieving innovation; it is a political and power medium that has to be shaped. It is seen as an asset to be fostered and funded, regulated, and restricted: Policies need to be developed, justified, and legitimized. The unrestricted or instantaneous production, diffusion, and use of new knowledge are no longer feasible—if they ever were. Today, the side effects and long-term impacts have to be taken into account. Risks and chances need to be identified as early as possible and public debates on side effects have

DOI: 10.4324/9781315387109-3

to be initiated. There is obviously a flip side, namely the growing body of non-knowledge, uncertainty, and ignorance that has become an issue for late-modern knowledge societies. Path decisions about which knowledge is societally desirable and acceptable have to be made in the early phases of knowledge development. To describe the shift in the relevance of scientific knowledge, Gernot Böhme and Nico Stehr (1986) have coined the terms "knowledge society" and "knowledge politics": Knowledge politics seems far more fundamental than traditional science and technology politics; scientific knowledge underlies the entire (ambivalent, Janus-faced) technological innovation process, and it shapes the future of societies at large, as Nico Stehr (2005) argues.[5] An earliness-oriented shaping of knowledge is the cornerstone for building society (Schmidt 2007b).

The notion of interdisciplinarity plays a central role in recent knowledge politics—as will be shown exemplarily in the following chapter. However, the term does not only convey ideology. It can also serve as an excellent door opener for a deeper (and more critical) engagement with the future of our knowledge society as a whole. Of particular relevance are the intersection and relation between science and society. Since this essentially implies transcending the boundaries of the academy, some scholars prefer the notion of transdisciplinarity, as discussed in the previous chapter. It has become obvious that real-world challenges—for instance, those posed by climate change, international terrorism, or the Covid-19 pandemic—cannot be met by the established framework of disciplinary fragmentation and separation. Interdisciplinarity seems to offer a promising way out. A "new mode of knowledge production" that appears to be both possible and indispensable is envisioned, as Michael Gibbons et al. (1994) maintain. Paul Forman (2012) argues in the same vein that post-industrial knowledge societies require interdisciplinarity in order to facilitate innovativeness and to ensure international competitiveness but that classical-modern industrial societies are based on a disciplinarily mode of highly specialized knowledge.[6] From a more sceptical point of view, interdisciplinarity can be interpreted as a sign of transformation towards a "post-academic type of knowledge," according to the analysis presented by John Ziman (2000). For Ziman, interdisciplinarity signifies the selling of the sciences to a global economy that eliminates any critical or reflexive viewpoint and thus is a danger to the paradigm of democratic fact-based decision making in society at large.[7] This view underlines again the inherent ambivalence—even dialectic—of inter- or transdisciplinarity. Let us now consider one particular case.

A prominent example of present-day knowledge politics with regard to interdisciplinarity is provided by the US National Science Foundation (NSF). The NSF explicitly refers to interdisciplinarity when it speaks of the "integration" or "convergence of nanotechnology, biotechnology, information technology, and cognitive science" (NBIC) (Roco and Bainbridge 2002; cf. Baird et al. 2004). Interdisciplinarity in this context is expected to guarantee "technological innovation" by a "synergistic combination of four major 'NBIC' provinces of science and technology."

The question addressed in this chapter is: To what extent can we legitimately label the NBIC vision "interdisciplinary"? Following the classification of four different types of interdisciplinarity proposed in the previous chapter, we can now critically inquire into the specific type of interdisciplinarity dominating the NBIC scenario. Reflection on interdisciplinarity could contribute to fostering reflection on and a review of knowledge politics. Thus, the *Philosophy of Interdisciplinarity* shares much with what Rouse (1987) once programmatically labelled a "political philosophy of science" and what Frodeman and Mitcham called a "philosophy of science policy" (Frodeman and Mitcham 2004).

The nanoresearch program

In the report on *Converging Technologies for Improving Human Performance* (Roco and Bainbridge 2002), the NSF advocates a specific type of interdisciplinarity, as will be shown. In general, the notion of interdisciplinarity is broadly present. The Roco–Bainbridge report, as it is now known, recognizes "specific needs to develop [...] interdisciplinary activities" (ibid.). To accomplish this task, the report recommends "interdisciplinary education programs" and "concentrated multidisciplinary research programs" (ibid., 11). Interdisciplinarity "means more than simply coordination of projects [...]. It is imperative to integrate what is happening" (ibid., 32). In addition, the report aims to promote "an interdisciplinary [and ...] holistic view of technology" (ibid., 13). A truly "interdisciplinary approach" "should focus on the holistic aspects and synergism" between different kinds of technologies (ibid., 16). So let us look in more detail at the line of thought presented in the Roco–Bainbridge report.

Diagnosis

The overall aim of the Roco–Bainbridge report to foster interdisciplinary collaboration and interdisciplinary integration in the field of engineering sciences (and beyond) is readily comprehensible. In particular, engineering sciences are largely a diverse patchwork made up of different branches such as electrical, mechanical, material, civil and military, environmental, material, information/computer, chemical, and biomedical engineering. Classical technologies are discipline-bounded or domain-restricted. They are developed and applied in specific contexts, such as biomedical technologies in the field of medicine or information technologies in the context of information processing, management, and storage (Schmidt 2004). Not only have the natural sciences and humanities been splintered into disciplines and subdisciplines through specialization and fragmentation, but so have the engineering sciences.

Although the overall focus of the Roco–Bainbridge report lies in "improving human performance," the report is essentially concerned with interdisciplinarity in engineering sciences and technology. During the last 70 years, a great deal of effort has been made to bring together the various areas of science-based technologies, including earlier attempts in the field of cybernetics and information theory in the 1940s and micro systems technology in the 1970s and 1980s. However, little overall progress has been made. Engineering sciences remain a patchwork. According to the Roco–Bainbridge report, the boundaries between engineering disciplines restrict the pace of technological innovation. "The traditional tool kit of engineering methods will be of limited utility in some of the most important areas of technological convergence" (Roco and Bainbridge 2002, 11). The objective, then, is to overcome the limitations of engineering sciences in order to ensure "technological superiority" (ibid., 14)—whatever that could mean in specific contexts.

Towards a new fundament

The basic intention when advancing and facilitating (various disciplines of) engineering sciences is to develop a non-disciplinary fundament, which the authors of the Roco–Bainbridge report assume underlies all engineering sciences. Until now, this fundament has been lacking. Present-day engineering sciences appear to be unstructured, heterogeneous, and diverse. The Roco–Bainbridge report presupposes that a common interdisciplinary fundament is in fact possible and desirable: The deeper this fundament and the stronger the related unification, the wider the positive impact it will have on the development of day-to-day technologies. The Roco–Bainbridge report can be seen as an attempt to facilitate an interdisciplinary foundation of engineering sciences on a deeper level.

Such a foundation will carry tremendous power for a novel kind of technological development and for innovation in general. The foundation is a technological one. These *foundational or fundamental technologies* can also be called *enabling technologies:*[8] They enable, create, and foster particular technical systems in applied branches and disciplinary domains. The way to promote engineering sciences is therefore to engage in an interdisciplinary convergence process of four central types of technologies. "The phrase 'converging technologies' refers to the synergistic combination of four major NBIC (nano-bio-info-cogno) provinces of science and technology, each of which is currently progressing at a rapid rate" (Roco and Bainbridge 2002, ix). This NBIC scenario, as it is called, lies at the very heart of the technoscientific research program of the Roco–Bainbridge report and the understanding of interdisciplinarity articulated therein. But what kind of fundament does it aim to provide? What are the underlying assumptions?

Underlying assumptions

The thesis that such a convergence is possible is based on ontological and epistemological assumptions. We find both assumptions throughout most of the report. For instance: "In the early decades of the 21st century, concentrated efforts can unify science based on the unity of nature, thereby advancing the combination of nanotechnology, biotechnology, information technology, and new technologies based on cognitive sciences" (Roco and Bainbridge 2002, ix). That is to say, the NBIC scenario—with its vision of interdisciplinary convergence—is rooted in the traditional ontological conviction of the unity of nature. A metaphysical naturalism pervades the Roco–Bainbridge report. From an ontological perspective, the epistemological assumption of the unification of science can be regarded as a mere derivation:

Unity of nature ⇒ unification of science ⇒ convergence of technologies.

In this vein, the report maintains that the time has come for "unifying sciences and converging technologies" (ibid., x). In fact, unification of the sciences, including (and particularly among) the engineering sciences, will provide the fundament for bringing all technologies together.

The visionaries of the NSF even foresee a more comprehensive unification that would include the social sciences and the humanities. According to the NSF, interdisciplinarity is not restricted to engineering and natural sciences but encompasses essentially all academic disciplines. The NSF Roco–Bainbridge report assumes that it is "possible to develop a predictive science of society" (ibid., 22) and a "unified cause-and-effect understanding of the [entire] world" (ibid., x). The report explicitly advocates a "predictive science of societal behavior" (ibid., 158ff)—a view that comes close to Auguste Comte's positivist program in the early 19th century. For the authors of the report, the cultural and social spheres are part of nature and are ruled by mathematical laws—a viewpoint that is indeed strongly naturalistic. They do not define nature solely as an object of study for the natural sciences. Their conceptualization of nature is certainly very broad; nature seems to be taken as a synonym for the nomological sphere in general, where causality determines the past and the future of the object. Therefore, cause-and-effect explanations become feasible. The report explicitly states that "a trend towards unifying knowledge by combining natural sciences, social sciences, and humanities using cause-and-effect explanation has already begun" (ibid., 13). Unification is feasible because causality or an all-encompassing nomological structure is given by nature—including social and cultural "nature" as part of nature overall.

By arguing in favour of interdisciplinary unification based on causal explanations, the Roco–Bainbridge report assumes that a certain kind of reductionism is a successful strategy: The tenor of the report is that scientific propositions should be reduced to cause-and-effect (or even mechanistic) explanations that make reference to general laws and the nomological structure of the world. Only such reductionism guarantees scientific and technological progress. Reductionism means, first, to establish relations between disciplines and, second, to find cause-and-effect relations. Thus, reductionism is regarded as both a tool and result of successful interdisciplinarity. According to the report, some partisans for independence of biology, psychology, and the social sciences have argued against 'reductionism', asserting that their fields had discovered autonomous truths that should not be reduced to the laws of other sciences. But such a discipline-centric outlook is self-defeating, because as this report makes clear, through recognizing their connections with each other, all the sciences can progress more effectively.

(ibid., 13)

The authors see no contradiction between interdisciplinarity and reductionism. Rather, we find here a non-disciplinary reductionism, which is a kind of reductionism typically referred to as "neutral reductionism." A reduction towards unified physics, evolutionary biology, or chemical engineering or to cognitive psychology or behavioural economics is not what the authors intend.

Technological reductionism

As discussed, the overall aim of the Roco–Bainbridge report is to foster a convergence. The unification of science, and the implied kind of interdisciplinarity, is regarded not as an end in itself but as a means to facilitate the convergence of the engineering sciences. Technologies, in particular nanotechnologies, are in the focus of the report.[9]

The basic goal of the report is to build capacities for a "technological mastering" and "monitoring" of the world. A novel kind of enabling technology is deemed necessary to construct, create, and control reality. In essence, technology is framed as a value-neutral instrument to accomplish different human ends. The instrumentalist view of technologies is also expressed in poem-like form in the report: "If the Cognitive Scientists can think it, the Nano people can build it, the Bio people can implement it, and the IT [information technology] people can monitor and control it" (ibid., 13). Thus, the IT people would control what the cognitive scientists think up. The presupposed causal nexus of the entire world, including associated control ambitions, is not perceived to contradict human action and free will. Such a futuristic vision resembles Laplace's deterministic worldview and the Demon he evoked in the early 19th century.

The NBIC scenario goes beyond the symmetry of nano, bio, info, and cogno. Nanotechnology seems to provide the fundamental material basis of enabling technologies. The nanoscale is where the interdisciplinary convergence of the four technologies takes place. According to the Roco–Bainbridge report, "convergence of diverse technologies is based on material unity at nanoscale and on technological integration from that scale. The building blocks of matter that are fundamental to all sciences originate at nanoscale" (ibid., x). Nano-objects are at the centre of the interdisciplinary convergence of technologies. Convergence is the pacemaker for the technoscientific, interdisciplinary process of unification, and unity is the ultimate point. The ultimate point is the point of total control, the point of Archimedes. The Roco–Bainbridge report presupposes the existence of unity on the level of objects—in other words, a techno-ontological unity in, and on the level of, nano-objects; convergence means the convergence of all engineering sciences on and in the nanotechnological objects. Top-down convergence is linked with bottom-up construction: "shaping the world atom-by-atom"—which is in fact the vision published by the US National Nanotechnology Initiative in 1999: The greater the convergence, the greater the power of the human engineer. Converging technologies are enabling technologies since they enable the development of technical systems in various branches of application.

In sum, the position expressed in the Roco–Bainbridge report can be called *technological reductionism* (Schmidt 2004). Interdisciplinary convergence is based on technological reductionism—and this kind of reductionism is merged (bottom-up) with technological constructivism or creationism (Kastenhofer and Schmidt 2011). As used here, the term reductionism denotes a reduction to techno-objects (ontological level) and not to theories, concepts, or knowledge (epistemological level) or to methods and procedures (methodological level). Apparently, the visionaries of technological reductionism assume that shaping the bottom level, or the nanocosm, implies an intentional shaping of the mesocosm, macrocosm, and megacosm. A linear determinism from the nanocosm to the other cosms seems to exist. In consequence, technological reductionists dismiss other scales of action in the world, such as the mesocosm, macrocosm, or megacosm. They do not regard such scales as being relevant for mastering and controlling the world.[10]

Technological humanism and the next industrial revolution?

The Roco–Bainbridge report advances a strong technological optimism. The reductionist program, in particular technological convergence, the report states, "could become the framework for human convergence. The twenty-first century could end in world peace, universal prosperity, and evolution to a higher level of compassion and accomplishment" (Roco and Bainbridge 2002, 6). The authors take it for granted that NBIC technologies will move the world dramatically in a positive direction. "Converging technologies could achieve a tremendous improvement in human abilities, societal outcomes, the nation's productivity, and the quality of life" (ibid., ix). "Therefore, the success of convergent technologies [...] is essential to the future of humanity" (ibid., 26). Convergence will promote activities to "improve human performance" (ibid., 26). According to the Roco–Bainbridge report, humans' social behaviour—and humanity—can be technologically shaped. Humanity can be fostered by technology—not just by education, cultivation, civilization, and socialization. Moreover, the human brain is thought to be open to technological perfection. The "technological humanism" presented in the report is portrayed as the cutting-edge kind of humanism that shares much with the transhumanist movement.

In addition to technological humanism, a number of visions with regard to the national economy and security are very significant in the Roco–Bainbridge report: "Technological superiority is the fundamental basis of the economic prosperity and national security of the United States, and continued progress in NBIC-technologies is an essential component for government agencies to accomplish their designated missions" (ibid., 14). The development of NBIC technologies is expected to ensure and promote the international competitiveness of the US economy and US firms.

Taking all the technological benefits for individuals and society into account, the report states that converging technologies can "initiate a new renaissance" and that a "next industrial revolution" will emerge soon (ibid., x). Since the era of the Renaissance with its historical phase of transition and transformation—from the medieval age to modern times—scientifically based technological advancement has been equated with human and societal progress. The NBIC scenario comes close to the traditional-modern view of technological progress and societal prosperity developed in the early 17th century. Such technological optimism can be traced back to the politician and philosopher Francis Bacon and his contemporaries. From Bacon's time on, science-based technologies were regarded as the pathway to a better future; they were not perceived as ambivalent. In line with this thinking, the Roco–Bainbridge report expresses the belief that technologies in themselves offer new and positive opportunities for society as a whole. According to the report, "science and engineering must offer society new visions of what it is possible to achieve through interdisciplinary research projects designed to promote technological convergence" (ibid., 14).

The first part of this chapter was concerned with disclosing the broader context and philosophical background of the Roco–Bainbridge report while also discussing the prominence of the political buzzword "interdisciplinarity." Yet the notion of interdisciplinarity as used in the report has not, up to this point, been fully elaborated or defined. I will therefore now go further and endeavour to clarify what kind of interdisciplinarity is involved in the NBIC scenario and promoted so strongly throughout the report.

Different types of interdisciplinarity

Using the framework of the four types of interdisciplinarity presented in the previous chapter, we can determine which one is the most dominant in the Roco–Bainbridge report. With which type of interdisciplinarity is technological reductionism associated? Plainly put, the report does not have much to offer with regard to theories and methods, and it contributes only marginal aspects with regard to pressing societal problems or issues.

A view on theories

Providing a coherent theory, or a consistent view of the things, is apparently not the aim of the Roco–Bainbridge report. A patchwork of models would, it seems from the report, work well if it provided a sufficient and efficient basis for technological innovations. Theories are not regarded as ends-in-themselves: rather, they are viewed as means and instruments. They are judged by the question of whether they actually contribute to the development of new technologies—or not. Theories are not per se superior to phenomenological knowledge or piecewise models. In short, technology, and not theory, is the aim; technological intervention is preferred over theoretical representation. On the other hand, the Roco–Bainbridge report concedes that theoretical knowledge is indispensable for facilitating the research and development of novel emerging technologies. The patchwork of present-day engineering sciences limits progress. In order to promote engineering science and to develop enabling technologies, we have to "integrate what is happening" (Roco and Bainbridge 2002, 32). Nothing is more practical than an adequate theory.

In fact, theoretical orientation for the sake of practical relevance makes the NBIC scenario an excellent example of a "technoscience."[11] Natural sciences and theories on the one hand and engineering sciences and technologies on the other hand are merged. A more detailed analysis could show that, because of its practical and pragmatic orientation, the Roco–Bainbridge report fosters only a weak understanding of theory in which a theory is not conceived of in the sense of a deductive-nomological type of explanation that is still the underlying objective of the paradigmatic unification project in disciplinary physics. Thus, the report is hesitant and prefers to talk about the integration of models, concepts, and knowledge rather than about theories.

If we assume for a moment—contrary to what has been elaborated so far—that theories were the aim of the NBIC scenario, is the NBIC scenario successful? The answer is: not at all. It is hard to identify any common theoretical umbrella or any interdisciplinary theory in the NBIC scenario, even if we consider only the weak understanding of theory. Admittedly, we do find progress in terms of theories within the disciplinary branches of (nano-)physics or chemistry, but the progress can hardly be considered interdisciplinary.

Are new methods involved?

The Roco–Bainbridge report does not aim to provide an interdisciplinary method and a unified methodology. Methods are regarded as instruments to obtain technical knowledge. What matters most are the efficiency and the effectiveness of methods in general. No reference is made to interdisciplinarity or integration of methods. If unification can help to increase efficiency, it is highly desired. However, the methods we find in the NBIC branch are not very interdisciplinary. They are based on advancements in the realm of physics; some can be traced back to chemistry and molecular biology. Not surprisingly, it was a physicist, Richard Feynman (2003), who gave the first programmatic speech on nanotechnology (in 1959). He declared that there seems to be "plenty of room at the bottom."

The NBIC technologies are fuelled mainly by methodological improvements in the field of physics. Nanotechnology owes its rise essentially to new physical instruments such as the scanning tunnelling microscope and the atomic force microscope. These stem from advanced developments in physics at the beginning of the 1980s. If the core of the NBIC scenario is rooted in nanotechnology, then it is rooted in physics. In fact, method disciplinarity is widely predominant.

Considering interdisciplinary problems and purposes

NBIC convergence can hardly be regarded as problem-oriented interdisciplinarity. Problem orientation requires at least a minimal consideration and justification of purposes or goals. Only very general and unspecific goals, such as human enhancement and fulfilling the basic needs of the less-developed countries, are formulated in the report. In contrast, a problem-oriented approach would involve the deliberate setting of goals and reflection on purposes—which encompasses the option to revise goals. Problem-oriented interdisciplinarity intends to focus on and solve societal and ethically relevant problems by explicitly reflecting on goals—in some cases by harnessing and developing new technologies; but technologies are just means that serve to reach ends. The Roco–Bainbridge report does not explicate or attempt to initiate a reflexive discourse about purposes. At the same time, the report conveys a certain fascination with the idea of technological development per se, interlaced with the vague idea of human enhancement: tapping opportunities but lacking orientation. For instance, the report shows no reservations with regard to broad military uses. The improvement of converging technologies for battlefield domination is envisioned. Thus, the Roco–Bainbridge report does not fit into the concept of problem-oriented interdisciplinarity.

Focusing on technical objects

Up to this point in the analysis, the findings derived in this subsection point to a negation of interdisciplinarity in the NSF report: There is a lack of theory-oriented and method-oriented interdisciplinarity and a very limited degree, if any, of problem-oriented interdisciplinarity. What can be said of object-oriented interdisciplinarity? According to the definition in the previous chapter, two different kinds of object-oriented interdisciplinarity have to be taken into account. (1) A strong version assumes that objects are time-invariantly located on boundaries due to the universal layers of reality (universal object interdisciplinarity). Pursuant to an underlying ontological realism, these objects were called interdisciplinary objects. (2) A weaker version concedes that the boundaries have not always existed and do not exist for ever. According to this kind of object-oriented interdisciplinarity, boundaries are constructed and thereby humans construct reality. Humans construct boundaries and create objects on the boundaries—or in short: (artefactual) boundary-objects (partial object-oriented interdisciplinarity).

Given this distinction, it seems obvious that the NBIC scenario is concerned with created and constructed nanotechno-objects which did not exist previously in nature, although they are based on the laws of nature (e.g., new materials, products, and processes). According to the Roco–Bainbridge report, the convergence of the four technologies is assumed to take place at the scale of the nanotechno-object and not in theories, methods, or problems: "Convergence of diverse technologies is based on material unity at nanoscale and on technological integration from that scale. The building blocks of matter are fundamental to all sciences" (Roco and Bainbridge 2002, ix). In the very small and real-constructed world of the nanocosm, everything seems to converge. From this perspective, the nanotechno-objects can be labelled "interdisciplinary."

Techno-objects are at the core of the heterogeneous and diverse fields covered by the umbrella term "nanotechnology," which includes fields such as electron-beam and ion-beam fabrication, molecular-beam epitaxy, nano-imprint lithography, projection electron microscopy, atom-by-atom manipulation, quantum-effect electronics, semiconductor technology, spintronics, and micro-electromechanical systems. In these examples, constructed and created interdisciplinary techno-objects are essential parts of present-day reality or of the reality to come ("ontological" type). Techno-objects are spreading and populating our world.

At this point, it is interesting to see how the real-constructed nano-objects relate to physics. On the one hand, nanotechno-objects belong to the domain of physics; they are located on boundaries between the quantum microcosm and the mesocosm. On the other hand, the Roco–Bainbridge report aims to produce instrumental knowledge about and serve enabling technologies, not to obtain true objective knowledge as in (traditional) modern physics. Although the boundaries between physics and engineering sciences are highly disputed, it is worth stressing that 'converging technologies' does not imply a convergence to objects belonging to disciplinary physics but rather a convergence to technoscientific nano-objects—which are objects for technological purposes. That is why we do not have in the NBIC scenario a reduction to disciplinary objects such as objects of physics but a reduction to interdisciplinary (real-constructed) objects. In this reductionism, nanotechno-objects are located at the boundaries between physics, chemistry, biology, and some engineering sciences. Put in the words of the physicist Richard Feynman (2003), "there is plenty of room at the bottom" for non-disciplinary nano-objects.

Critique

An interesting question concerns whether the above-envisioned techno-object–centred type of interdisciplinarity is based on sound arguments. This is not the case, as the *Philosophy of Interdisciplinarity* can disclose. It is hard to see how research programs such as those advocated in the Roco–Bainbridge report could actually be successful.

The report overestimates the feasibility of technological reductionism. Major limits are conspicuous. The reductionist approach might be effective for the development of specific materials or certain processes, such as super-conductivity or quantum computing, but not in general as claimed in the report. The thesis that nature—and technical systems—can be created and constructed atom-by-atom is misleading: The constraints of physics and chemistry are too severe.

The following line of argument refuting technological reductionism is advanced: Since NBIC visionaries recognize complex systems theory and related approaches, including theories of self-organization, nonlin-ear dynamics, dynamical systems theory, chaos theory, and synergetics, as being one of their fundamental theories—which is what they claim in their report—they should be aware of the limits of reductionist strategies and hence of the methodological limits of technical manipulation. In more detail, one important lesson of complex systems theories for all mathemat-ical and engineering sciences known today is the fundamental role of non-linearity and instability.[12] Since nature is governed by nonlinearity, physical objects and technical systems can be structurally and dynamically unsta-ble. Flipping points, bifurcations, and chaos can occur. Small changes in initial or boundary conditions produce large effects in the overall system properties and functionalities. Such phenomena are called butterfly effects. According to M.L. Roukes, a physicist engaged in the field of nanosystems, instability challenges the nanotechnological bottom-up strategy and limits technological control of the tiny objects. There is no direct path of control from the nanocosm to the macrocosm: The smaller the objects are, the more unstable they can behave and the more the nano-effects are amplified into the mesocosm without any control; perturbations on the nanoscale cannot be handled and controlled in all relevant details. "The instability may pose a real disadvantage for various types of futuristic electro-mechanical sig-nal-processing application" (Roukes 2001, 37).[13]

In consequence, two intermingled limitations need to be taken into account in the nanotechnological bottom-up strategy, as Richard Smalley (2001) points out. Smalley calls one the "fat finger problem" and the other the "sticky finger problem." (a) Because the fingers of a manipulator arm or technical apparatus themselves must be made of atoms, they have a certain irreducible finite size. In many applications, the manipulating fingers are far too "fat." (b) Furthermore, the atoms to be shaped or rearranged by nanotechnology will be too "sticky": The atoms of the manipulator arm will adhere to the atom of the nanostructure being moved. It will often be impossible to release this nanostructure in precisely the right spot. So there is no isolation and no definite border between the nano-object to be shaped and its surroundings. A kind of holism emerges in the nanocosm—based on nonlinearities and instabilities in the classical as well as in the quantum regime.

Briefly stated, complex systems and self-organization theories call into question (a) the experimental repeatability and technical (re-)producibility and hence the technical manipulation in general of technical structures, features, and properties and (b) their anticipability and predictability. (c) The model- or theory-based knowledge is restricted since nonlinear systems are effectively irreducible to a single comprehensible model. A cause-and-effect explanation of the entire physical and technical worlds, as the Roco–Bainbridge report envisions, is impossible.

Such limitations challenge reductionist strategies of explanation and of intervention and manipulation. Technical design, creation, and construction are limited. Therefore, the visions of the NBIC program—that NBIC convergence based on nanotechno-object-oriented interdisciplinarity is capable of shaping and manipulating the world atom-by-atom—are not based on sound arguments.

Summary and prospect

Inter- and transdisciplinarity are political terms: They show up as key notions in knowledge politics and are constitutive for present-day political epistemology. The *Philosophy of Interdisciplinarity* conceptualized here investigates the background of recent knowledge politics. This approach can be linked to what Rouse (1987) called "political philosophy of science" and to Frodeman's and Mitcham's (2004) "philosophy of science policy." The intention behind such an analysis is to open up avenues for public consideration and awareness of the eminent politicity, power, and authority of technoscientific knowledge. The notion of "knowledge politics" underlines that decisions pertaining to knowledge production determine the society of the future.

Based on the clarification and classification undertaken in the previous chapter, this chapter has presented a case study analysing one of the most influential programs in US knowledge politics and probably worldwide: the Roco–Bainbridge report on converging technologies. It has shown that techno-object–oriented interdisciplinarity is predominant in the Roco–Bainbridge approach. We can identify an "ontopolitics"—in line with David Chandler's (2018) approach.

The identification of different types of interdisciplinarity provides a framework for distinguishing different kinds of knowledge politics pursued by research programs. Understanding interdisciplinarity turns out to be a cornerstone for an analysis and assessment of a specific knowledge politics. That is to say, reflection on interdisciplinarity is the very basis for a normative review and a (potential) revision of recent knowledge production—one of the major policy fields in our late-modern societies. We need a broad discourse on what is worth knowing and for what purpose—and a discourse of this magnitude has to be regarded as one that is inherently political. Recognizing the politicity of interdisciplinarity can be considered central to what Ulrich Beck and Christoph Lau (2004) called *reflexive society*. Shaping interdisciplinarity needs to be regarded as a cornerstone for building late-modern reflexive societies.

Notes

1 See Maasen (2010), Bogner et al. (2010), and Schmidt (2007b). The term "politics," as used here, indicates that power aspects, authority, institutions, and regulations are always involved in interdisciplinarity. This point goes beyond the dichotomy that is prominent in phrases such as "speaking truth to power."

2 This term "political epistemology" is introduced with different meanings and connotations; see, for instance, Omodeo (2019).

3 Weingart (2000) sees "innovation" as the central characteristic of interdisciplinarity: "Interdisciplinarity is not the promise of ultimate unity, but of innovation" (ibid., 41).

4 This point of critique is advanced by Böhme and Stehr (1986), Stehr (2005), and also Carrier (2011) and Wilholt (2012).

5 Knowledge politics has been emerging ever since it became apparent that scientific knowledge throughout the innovation process proves to be far more complex than in the past: That is to say, time-consuming, wide-ranging, risky, expensive, sensitive to governmental regulations, dependent on public and private funding, and so on.

6 Forman (2012) argues that "modernity entailed disciplinarity," whereas "postmodernity entails antidisciplinarity." Although I do not use the same words as Forman ("postmodernity," "antidisciplinarity"), I share his analysis.

7 For Ziman (2000), disciplinary sciences, in their procedural search for objectivity and truth, could serve as a paradigmatic form of and blueprint for democracy at large. Decision-making in science as well as in society should be based on rational argumentation—and not on the power game of diverse interest groups (ibid., 153). Ziman's position implicitly shares core elements of Habermas's discourse theory (Habermas 1993).

8 The term is elaborated by Schmidt (2004, 39/42); see also Baird et al. (2004, 13).

9 Cognitive sciences, for instance, are mainly understood and defined from the perspective of new technologies; the NBIC report refers to "new technologies based on cognitive sciences" and to "neuro-technologies" (Roco and Bainbridge 2002, x).

10 According to this position, the mesocosm, macrocosm or megacosm do not seem to possess their own distinct ("supervenient") properties. This is, of course, a strong claim. It is based on a straight naturalistic viewpoint that is based on the classical conviction of a continuous cause-and-effect nexus of the world, and especially a naturalistic line from the nanocosm to the macrocosm. The phrase "shaping the world atom-by-atom" neglects the relevance of engineering sciences on scales of the microcosm, mesocosm, macrocosm or megacosm; it focuses only on the nanocosm.

11 See the subsequent chapter.

12 See Chapter 7 in more detail.

13 A further limitation is given by the fundamental threshold for minimum operating power: The random thermal vibrations and fluctuations of a technical device impose a "noise floor" below which real signals become hard to discern.

4 History and technoscience
Tracing the historical roots of object-oriented interdisciplinarity

Instrumentalist mindset

The recent debate on interdisciplinarity is dominated by an instrumentalist mindset—as exemplarily seen in the previous chapter with regard to the converging technology program and its object-oriented interdisciplinarity (Roco and Bainbridge 2002). Interdisciplinarity appears to be a means for dealing with or constructing interdisciplinary technical objects, artefacts, devices, and machines. But what ends or purposes does it serve?[1]

Instrumentalists and strategically oriented researchers hardly see it as part of their research activity to explicitly reflect on problems, goals, and purposes. They accept given goals if those goals match their research interests or are pursued by funding agencies, institutions, or corporations; they start their research from here on in. Specifically, they do not reflect on the role and function of their research in society. The rash acceptance of given goals is typical of instrumentalist approaches in diverse branches, no matter whether the proponents are engaged in military research (Fuller 2017) or committed to sustainability studies or transformative research. Although instrumentalists are aware that goals can stem from different sources— namely from an intra-scientific, societal, or economic context—and also that pursuing some goals might be preferable to pursuing others, they do not deem it part of their research practice to consider goals explicitly. Reflection on or even critiquing and (potentially) revising goals seem beyond the scope of their professional competencies as researchers. In other words, they accept a division of work and confine themselves to means while maintaining that others, such as politicians, university leaders, or directors of funding agencies, are responsible and accountable for their respective goals.

Although, or because, instrumentalists accept externally given goals, they paint a value-free picture of research and interdisciplinary activity: They see science as value-free and technology as neutral. The interplay between science and research on the one hand and society, politics, and the field of economy on the other hand is conceptualized by a simplistic, dualistic input–export model. In sum, their conceptual framing of interdisciplinarity is based on objects, methods, or theories—regardless of whether they work in fields dealing with cosmology, the evolutionary history of the earth, synthetic bio-materials,

DOI: 10.4324/9781315387109-4

nano-optics, supply chain management, innovation studies, military research, risk management, sustainability studies, or global change policy consultancy. According to this decisionist viewpoint, it is up to the ethicists (and to scholars of the normative disciplines such as jurisprudence) to judge research goals and up to the politicians or legal bodies to make the decision on research agendas.

The instrumentalist position is most prominent in what I have termed object-oriented interdisciplinarity, although it is also present in programs and projects subscribing to the method- or theory-oriented type of interdisciplinarity—and in the many projects involving problem-oriented interdisciplinarity, as we will see later on. The *techno*-object orientation discussed in the previous chapter is a particularly standout example of the instrumentalist viewpoint. Technical objects, such as nano-particles, nuclear power plants, machine tools, fabrication machines, oncomice, personal computers, artificial intelligence systems, skyscrapers, water supply systems, chemical weapons, or mobility systems but also buildings, bridges, and bikes, can be regarded as being interdisciplinary in their core. The aim of this chapter is to provide clarification: In what way and to what degree can such objects be considered interdisciplinary? What is the ontological basis—in brief, the ontology—of these objects and of the related "ontopolitics"?[2]

A look at the history of science shows that, although the term "interdisciplinarity" is seldom mentioned, most of the central characteristics attributed to the object-oriented type of interdisciplinarity are hardly novel—even though the objects themselves are often novel. In fact, these ontological characteristics have a fairly long tradition. They date back to Francis Bacon and his programmatic foundation of science in the early 17th century (Schmidt 2011b). Bacon's program of science as research—interlaced with a straightforward and strategic organization of the academy based on a disciplinary division of work for the purpose of producing new and discipline-transcending techno-objects—can serve as a paradigm for any kind of instrumentalist stance. Referring to Bacon helps us to address the ontology of these techno-objects. My particular diagnosis—that the recent interdisciplinarity hype is dominated by a Baconian instrumentalist viewpoint, including a strong focus on interdisciplinary objects—is therefore to be understood as a first step in overcoming or complementing the prevalent instrumentalist perspective.

No man's land of techno-objects

Let us look at the created objects, artefacts, and machines in more detail: Innumerable techno-objects and human-made artefacts populate our laboratories and our life-world. In the past and framed from the perspective of predominantly philosophical approaches, these created (mostly interdisciplinary) techno-objects remained hidden and invisible. For a long time, philosophers of science omitted to take the constructedness and createdness of our science-based reality into account. The philosophic locus of these discipline-transcending techno-objects of our socio-technological life-world was an ontological and epistemological no man's land.

The disregard for the technical dimension of our life-world has historical roots which are interlaced with a traditional understanding of science and, in particular, of scientific knowledge. Since the Ancient Greeks, and in a renewed form since the beginning of the modern age, scientific knowledge and insight had been reduced to a cognitive, theoretical, and contemplative dimension. If anything, technical objects and material instruments were seen as mere external servants for refuting or confirming scientific propositions; as means, they were marginalized. Even positions professing to relate to "reality" and not just to knowledge stayed in the realm of the cognitive-contemplative. This reduction served to detach science from human (technical) action and from (technologized) society and hence led to the mythologization of science.

The lack of consideration and reflection has given rise to a wave of critique.[3] After all, the "mind has left the vat" and "the brain the tank," as the French philosopher and sociologist Bruno Latour (1999) maintained: We live in a material(ist) culture. From the printing press to the computer, from gunpowder to nanobots to machine learning algorithms—material manifestations of humans are everywhere. However, philosophers, according to Ian Hacking (1983, 149f), "constantly discuss theories and representations of reality, but say almost nothing about experiment, technology, or the use of knowledge to alter the world." Scientific knowledge was considered to be *theoretical* knowledge and *cognitive-conceptual* representation; *theoria*, not technology, not technical *poiesis* or *praxis*; mind, not matter, as John Dewey (1929) and the pragmatist traditions critique. The ontology of the objects was not part of philosophic reflection. From this perspective, our understanding of the scientific enterprise and of our material culture has "never been modern," Latour (1993) laments. Other prominent lines of the critique can be traced back to Marx and Engels, who observed that philosophers have primarily interpreted the world; the important point, however, is to intervene in it or to change it. Such a line of thought should not be regarded as outdated or erroneous, although the revolutionary attitude has largely disappeared today.

What undoubtedly remains a central point is the need for an in-depth consideration of the material entities and technical artefacts in the process of knowledge production—and specifically in the related concepts of inter- and transdisciplinarity. Indeed, today's nano- or biosynthetic objects, for instance, renew the need for a materialistically well-informed ontology to render these objects perceivable and also open to critique. In order to contribute to a deeper reflection on discipline-transcending techno-objects, it is rewarding to go back to the origin of the project of modernity and to its founding father: Francis Bacon. Bacon's program is most prevalent and powerful throughout what has been called techno-object–oriented interdisciplinarity.[4]

Although Bacon is a well-known figure, his contribution to the present-day ontology and epistemology has seldom been discussed in the philosophy of science. Admittedly, this could be due to his approach at first glance appearing to be so unelaborated that philosophers have had little to say about it. Compared with his contemporaries like Descartes and others, Bacon seems to lack analytical clarity and distinctiveness. Often, he is seen as merely a politician and not as a philosopher. Yet ignoring Bacon proves detrimental to an analysis of the ontology of interdisciplinary techno-objects. Bacon is a key figure who could help us realize that a materialist or material-focused constructivist or constructionist turn is required—to address not only untamed nature but also the roughness and harshness of everyday man-made nature and of the constructed technical systems. It follows that, if we regard constructed objects as the driving force of modern and late-modern societies, we need to reconsider Bacon's idea of conceiving knowledge as a form of material-based power.[5] Such a view opens up avenues for considering that "all questions of epistemology are also questions of social order": We are aiming at the "*politics of things*, not at the bygone dispute about whether or not words refer to the world" (Latour 1999, 3f). This point has, of course, been made before by the *Frankfurt School* of critical theory: "Social critique, in particular critique of society is critique of knowledge, and vice versa."[6] Appreciation of Baconianism can provide an excellent fundament for a critical review of recent techno-object–oriented interdisciplinarity in order to focus attention on the ambivalence of modernity, on the latter's "politics of things" (Latour 1999), on "ontopolitics" (Chandler 2018), or on "technoscientific politics" (Hackett et al. 2008).

In the realm of Bacon's concept, constructivity and reality are not an antagonism but two sides of the same coin. What is constructed and created is reality; what we know depends on the artefacts we create, which in turn become real facts about the world we live in. Bacon's materialist view could tentatively be called *real-constructivism*. Rather than being described as realist or constructivist, empiricist or rationalist, Bacon's position can best be understood as real-constructivist since it challenges modern dichotomies, including those between constructivism and realism and between epistemology and ontology (Schmidt 2011b). An ontology of this kind can serve as a basis for a critical public and scientific discourse on techno-object–oriented interdisciplinarity.

Bacon and the roots of techno-object–oriented interdisciplinarity

Bacon is one of the founding fathers and central watershed figures of the modern era and his presence is ubiquitous in technoscientific objects in our lifeworld—even if it is debatable whether Bacon is to be seen as a product of his time or more as a promoter and proponent of a new era (Zittel et al. 2008). If experimentally based interventions in nature and if technical constructions and creations are deemed central to enabling the progress of knowledge for the benefit of society at large, Bacon is *the* epochal threshold person—not Descartes or Galileo, nor Leonardo da Vinci or other Renaissance engineers. It was Bacon who argued that science and technology, or knowledge and action, are twins—a position that is quite commonplace in the recent diagnosis of the so-called era of technosciences and underlies the object-focus of the prevalent type of interdisciplinarity.

Bacon was seen as marking the crystallization point—and he became the target of severe criticism—in respect of the negative side effects of science, enlightenment, and modernity in general: as Baconianism. Bacon's intellectual impact on the modern age is documented by an extensive history of references stretching from Boyle, Hooke, Newton, Diderot and d'Alembert to Kant and Hegel, Dewey and Peirce, Marxists and Pragmatists, and Liebig and Cassirer, right up to Hacking and the New Experimentalists. Twentieth-century philosophers who cite Bacon include Horkheimer and Adorno, Jonas and Passmore, Farrington and Bloch, Böhme and Meyer-Abich, Merchant and Fox Keller, and Schäfer and Krohn. Most notably, Kant ascribed the "revolution in ways of thinking"[7] to none other than Bacon in the preface to his *Critique of Pure Reason*. In Kant's famous words, Bacon opened up "the highway of science."[8]

That is hardly an exaggeration. Bacon's claim at the beginning of the 17th century was not unpresumptuous. It involved not only verbal, cognitive, or theoretical knowledge but also intervening and action knowledge, interlaced with the act of creating objects, constructing artefacts, and designing nature; not only the cognitive-contemplative interpretation of a given reality but moreover the altering, shaping, and constructing of nature and thereby of the world—in precisely the manner portrayed in the famous Feuerbach thesis by Marx. With Bacon, the idea of a technoscience-based transformation of a transform*able* and construct*able* world emerged, reaching out in time and accessing the future: Man as *homo faber* makes history; a *regnum hominis* is fostered by a *scientia activa*. Bacon's stance as presented in his *Novum Organon* (1620)—the main work of his famous but unfinished *Instauratio Magna*, the *Great Renewal of the Sciences*—reflects the pathos of an epoch breaker: *there* was the *old* Organon of Aristotle and *here* is a *new*, viz. his own Organon; *there* was the sunken Atlantis of Plato and *here* is his *New* Atlantis.[9]

Bacon's programmatic engagement for a historical rupture leads to his not being able to adopt in a comparative manner the "guise of a judge" like the "ancients," but that "of a guide" (NO I: Aph. 32). His objective is "to open up a completely new route for the intellect, one unknown and untried by the ancients" (NO I: 57). He assumes that an "experimental philosophy" (later called "science") *for* the future is both possible and necessary. Thitherto, however, "progress in the sciences" has been held back "by reverence for antiquity, for the authority of those held to be philosophy's great men and then by giving their consent to all that" (NO I: Aph. 84). Only the characters of "master and pupil" and "not of discoverer and improver of discoveries" have been brought forth (NO I: 15). Bacon extends his criticism of "human authorities" of "antiquity" in a way that can be regarded as a preliminary form of critique of ideology (NO I: Aph. 84): In his famous doctrine of "idols," Bacon shows how uncertain notions, false judgments, and circular thoughts arise when authorities are blindly followed without question (NO I: Aph. 38ff).

Instead of persisting in ideological thinking, Bacon points out the active skills, knowledge, and insights gained by craftsmen, doctors, and seamen through their (inter-)action with their environment and the world. Traditionally, technical practice and the creation of objects were devalued; they were deemed theoretically insignificant (cp. Dewey 1929). By contrast, Bacon sees the making, constructing, and creating as fundamental pursuits, which are at the same time worthy and capable of being improved. Technical capability therefore should be paired with theoretical knowledge, and mechanics with physics. Bacon's aims are to found, foster, and facilitate sciences and to make them useful to society at large. Such a perception comes close to some of the ideals behind the notion of "transdisciplinarity"—which also intersects with aspects of object-oriented interdisciplinarity.[10]

It is debatable whether Bacon does justice to his claim of being a guide into the "open sea" of fruitful discoveries, useful inventions, and novel objects. Bacon's ontology could be described under the headings of (first) motive and approach; (second) practice, method, and genesis; (third) manifestation, technical works, and the truth of artefacts; and (fourth) matter, material, and "nature." It will be illuminating to expound these four aspects, which are characteristic for the instrumentalist account in today's object-oriented interdisciplinarity.

Aim and motive inherent in Bacon's concept

For Bacon, knowledge is not just an end in itself but is above all a means to an end: *for* further knowledge, *for* a deeper truth, *for* better instruments, *for* a more efficient control or even creation of nature, and even *for* a *regnum hominis*. Bacon opens science to society—this is very much in line with the recent discourse on transdisciplinarity. According to Bacon, "the true and legitimate end of the sciences is nothing other than to supply human life with new discoveries and resources" (NO I: Aph. 81) that "may, to some degree, subdue and mitigate their needs and miseries" (NO I: 37). Bacon believes that the new, active science is positively utilizable in its very core and could be beneficial to all. Therefore, "the end [...] for this science is not the discovery of arguments but of arts," namely of technology and technical artefacts (NO I: 29).

The different objectives of the Greek academy and the new science in the form of research lead, Bacon holds, to "different effects. For *one* aims to beat an opponent in debate; the *other* to bend nature to work" (NO I: 29), the latter in order "to command [the] things" (NO I: Aph. 29). However, Bacon is not given to non-theoretical tinkering with direct utility, which he identifies in Leonardo and the artist-engineers of the Renaissance. He believes they went about their work aimlessly, governed by "hazard," randomness, and trial-and-error rather than by methods (NO I: Aph. 8). The "fruit-bearing knowledge" at which Bacon is aiming presupposes (and is interlaced with) "light-bearing" knowledge, meaning a "discovery of causes" (NO I: Aph. 99). The latter is necessary in order to learn and advance "the art of discovering" (NO I: Aph. 130) in the sense of "to discover something to enable everything else to be rapidly discovered by means of it": an *ars inveniendi* (NO I: Aph. 129). This method-facilitating *meta* discovery program—the discovery of the logic of discovery—may be regarded as the core of Bacon's approach.

If the source of societal progress lies in the investigation of nature for human needs, as Bacon believes, society has an interest and a stake in science. An active science requires division of work and a strong institutionalization—that is to say, suitable overall conditions for good scientific practice that traditionally were associated with a disciplinary structure of the academy.[11] "For only then men begin to know their own strength, when instead of countless men doing the same thing, some will be responsible for some things, others for other things" (NO I: Aph. 113). In his utopian narrative "New Atlantis," Bacon delivers a programmatic account of interdisciplinary cooperation based on division of work, similar to that observed in early capitalist craft workshops. "Salomon's House"—the scientific institution of New Atlantis—is not a subordinate authority of the state "Bensalem"; it is more an autonomous institution seeking to incorporate societal values in order to enable and secure human progress. Scientists are free to decide what is worth knowing if they subscribe to the betterment of human conditions. Such institutional freedom guarantees light-bearing knowledge and consequently, as Bacon sees it, the most fruit-bearing knowledge. What Bacon refers to as "fruit-bearing knowledge" is characteristic of an instrumentalist view of science.

To summarize, Bacon does not stylize science and research as being without interest or value-free. According to Bacon, science aims at light-bearing and fruit-bearing utility as a way of changing, shaping, and manipulating *given* reality and constructively creating *new* realities. This point of view is indeed very prominent in today's technoscience—and it also runs through the instrumentalist mindset that is broadly present in techno-object–oriented interdisciplinarity.

Practice, method, and genesis

For Bacon, intervention, construction, and creation are not just ends but also the means for obtaining novel knowledge as a basis for creating new and better technical objects. The methodological core of scientific investigation is the experiment: The focus should be on experimental science and not just on mere (philosophical) debating and arguing. Knowledge is attained neither by random trial-and-error nor by passive observation or pure thinking but by a systematic process of experimental production and technical activity. Man "does and understands only as much as he has observed by work [in nature...]; beyond this he has neither knowledge nor power" (NO I: 45). Bacon does not wish to

> put together a history of nature free and unconstrained (when, that is, it goes its own way and does its own work [...]) but much more of nature

restrained and vexed, namely when it is forced from its own conditions by human agency, and squeezed and molded.

(NO I: 39)

The most appropriate practical action for learning about and from nature therefore seems to be experimental intervention or, more specifically, construction and creation of an artificial nature; the latter is still nature since we cannot escape nature. The experiment is ontologically appropriate to nature because given nature tends *per se* to be taciturn and to guard its secrets within. According to Bacon, a passive observer's perspective, which was later also criticized in the pragmatist tradition by John Dewey (1929), is impossible: Achieving knowledge is always an artefact-based operation on things and hence an *act*. Facts are rooted in actions, in *facere*, which is the Latin word for "make."[12]

What argument does Bacon produce in support of the technical experiment? His argumentation is clearly anti-sensualistic: The sense "fails us"; it "deserts" and may "deceive" us (NO I: 33). For instance, if the "minuteness of [a body's] parts" or "its swiftness or slowness" predominates (NO I: 33), the human senses are overtaxed, making a transition from the "incommensurable to the commensurable" infeasible (NO II: Aph. 8). The idea that it is possible to refine the senses by technical instruments (e.g., by means of a microscope) could arise. However, that only shifts the problem—it is not sufficient. Only the experiment can provide the answer. The experiment leads away from the observer's to the actor's perspective of knowledge. It enables, under artificial/technical conditions, control of the boundary conditions. The latter may be varied to ensure reproducibility, namely independence from spatial, temporal, and subjective moments. Stability and thereby regularity are established by human action.[13] Consequently, Bacon "set[s] little store by the immediate and peculiar perception of the sense, but carr[ies] the matter to the point where the senses judge only the experiment whereas the experiment judges the thing" (NO I: 35). Bacon is therefore not a sensualistic empiricist—"the sense" is *not* the "measure of things" (NO I: 35).

So far, the methodological foundation has been laid without yet having established how knowledge is actually produced. In fact, the experiment is also central to Bacon's "inductive method" (NO I: 31). According to that inductive method, it is possible and necessary to "abstract both notions and axioms from things by a surer and firmer way" (NO I: 31/Aph. 18/19). However, these "things" should not be understood in a naïve-realistic way, namely as being simply given. The "thing" itself becomes accessible, or is even constructed, as something "new" in (or by) the experiment. In the 1980s, Andrew Pickering writes in a similar vein about the microphysical "construction of quarks" and an experimental "production of the world." Thus, the experiment involves not only intervention but also specifically construction and production (Pickering 1984). Bacon tells us we should not pass from the singular to the universal in a single step. Rather, it is important "to educe axioms successively and step by step" (NO I: 31). This type of induction ("exclusion or elimination theory of induction") includes a certain empirically based falsificationism and proceeds by way of "exclusions and rejections" (NO II: Aph. 18).[14] In Bacon's view, "men are allowed only to proceed by *Negatives* at first" in order to arrive at sound knowledge "after making every sort of exclusion" (NO II: Aph. 15). So "every contradictory instance wrecks a conjecture" (NO II: Aph. 18). The term "inductive method" therefore signifies a tentative interplay between experimental constructions, inductions, and deductions. In this way, broader statements can be obtained successively. By this method, science is "driven on, as it were, by a machine."

Bacon's accounts led some philosophers in the 19th century, such as John Stuart Mill, to misleadingly discern in them only a contribution to inductive *logic*—and not to inductive *methodology* or *technology*. Hans Heussler (1889, ii) soon refuted Mill's position in his comprehensive work on Bacon, in which he argued against giving Bacon a "one-sided empiristically biased label." Heussler's rationalist line of interpretation could be supplemented with an interventionalist one or, more specifically, with a constructivist or creationist one.

The inappropriateness of such contractions is also substantiated by Bacon's prominent example of the bees, in which he contradicts both the Renaissance engineers and pure empiricists *and* the rationalists and philosophers he perceives to typically stand in the Greek tradition. Bacon likens experimentalists and empiricists to "ants" because they only "store up and use things"; they merely gather data mindlessly. Rationalists, on the other hand, are like "spiders" that "spin webs from their own entrails"—without referring to the outer world; they spin empty theories and meaningless propositions. Bacon's ideal is the bee, which "takes the middle path." The bee "collects its material from the flowers of fields and garden, but its special gift is to convert and digest it" (NO I: Aph. 95).

With that, a methodological kernel of real constructivism has been articulated: the experiment as mediation between empirical realism and active/experimental constructivism. Reality—in other words, nature—also encompasses that which is constructed and created. Gaining insight always involves investigation and intervention; it is inseparable from action.

Material manifestation, technical works, and the truth of artefacts

Bacon's philosophy of the creation and construction of novel things is based on the advancement of knowledge, not on the systematics of what is known. The truth of knowledge—and of what was later called the context of justification in the 20th century—does not lie in its foundational anchoring, but in its being continually built up and enriched. Knowledge is knowledge insofar as it is progressing, developing, and advancing. Bacon denies any kind of atemporal truth. He sees truth as being relative, in the sense of "daughter not of authority but of time" (NO I: Aph. 84). Bacon speaks of "scientific advance" and "progress" in a way that will become typical for the modern era.

What then is *truth*—or true knowledge—for Bacon? In Bacon's account, truth is, in a double sense, not without works, machinery, or techno-objects. True knowledge is, in the *first* place, the knowledge that is brought forth on works—via technologically based instruments and experiments—using the method of induction; it is then presented in tables of laws. Because these artefacts are essentially based on the laws of nature, the experimentally produced facts relate to nothing else but nature. *Second*, truth manifests itself in works and techno-objects. Created objects are, as Bacon writes, "guarantors of the truth"; they designate materialized objectivity (NO I: Aph. 124). In sum, works therefore serve a dual purpose with respect to what can be regarded as truth; they play a role in both the genesis of truth and the manifestation of truth.

However, works do not only benefit the truth. The reverse is also valid: Truth serves as "new pledges of works" (NO I: Aph. 81). If technical systems fail, it is due to "ignorance of causes," to untrue knowledge (NO I: 45/Aph. 3): Truth is conducive to the production of a work or machinery. For Bacon, this is not a circular process but an iterative, self-dynamizing process of technical work and truth generation. Works and truth therefore are closely related. Their spearheads are pointed at *phenomenally* given nature, which has to be "conquered," "bent," and "constrained" (NO I: 29/Aph. 3). "Experience" and "judgment" should be "drawn from [...] the very innards of nature" (NO I: 33). Although the commonplace phrase ascribed to Bacon, *knowledge is power*, is not to be found in any passage of his writings, kindred definitions are indeed present. For example, he states that "those twin objectives, human Knowledge and Power, do in fact come together," that "truth and utility are here [...] the very same things," and that "human knowledge and power come to the same thing" (NO I: 45; Aph. 124; Aph. 3). His words are certainly to be understood not only in a descriptive sense but also in a normative sense. Only that which is based on power over nature should be considered true knowledge—and hence what is the case. The connection is first expressed as a *negation*: Without having knowledge about nature—that is, if the cause–effect relationships are "not known"—the "effect" will be missed and cannot be predicted; practical use for human purposes is lacking. Put in *positive* terms: "that which in thought [= knowledge] is equivalent to a cause, is in operation equivalent to a rule" (NO I: Aph. 3). Vico later provided a famous formulation: *verum et factum convertuntur*.

The term "rule" as found in Bacon's writings might erroneously suggest a Humean kind of rule-following in the sense of a nomological succession of events—or a Newtonian trajectory of an oscillating body. That is not what Bacon means. For Bacon, the word "rule" signifies an action rule or instruction: Whenever I wish to bring about an effect Y, I must do X. In this respect, Bacon sounds, to some degree, like an action theorist, such as Georg Henrik von Wright (1971). Von Wright distinguishes between doing and bringing about or causing. *Doing* represents the specific action of man, viz. setting initial conditions and causes, thereby inducing the regular succession of events in nature, whereas the *brought about* event or the intended effect is induced causally by the laws of nature—via a trajectory based on Newtonian mechanics. In order to act intentionally in nature, knowledge of the laws or of the nomological Newtonian mechanism of nature is necessary. Thus, knowledge of nature becomes central and foundational to knowledge of action and production. *Intervention* and *representation* are mutually dependent, a position that today is held by Ian Hacking (1983).

That mindset could have confounded traditional, religious lines of argument. Nature in the hands of man—is that not hubris? On the contrary, Bacon says. According to Bacon, the intervention dimension of knowledge is motivated and justified even from a religious perspective. Bacon speaks of God's "first fruits of creation" (NO I: 45) and goes on to encourage the "imitation of God's works" by the action of man: as a *second instance of creation*. "Discoveries are also like new creations" (NO I: Aph. 129). In Bacon's thinking, man is even assigned the task to "write a revelation and true vision of the Creator's footprints and impressions upon His creatures" (NO I: 45). The religious motives of the modern era are hardly to be overestimated.

In summary, real-constructivism in terms of knowledge means the progressive manifestation of truth *in* works and machinery, thus *through* real constructs and technical things: that is, "thing knowledge"—a term coined by David Baird (2004). Insofar as real constructs are useful, they are based on truth.

The core of constructed and created objects: nature as mathematical law

Such an understanding of truth incorporates a specific conception of nature. Bacon repeatedly rejects large systems of natural philosophical speculations and metaphysics. But even he cannot dispense with a pre-understanding of "nature" that includes several assumptions. A certain degree of metaphysics is indispensable—even for those wishing to evade metaphysical questions.

Bacon rejects the Aristotelian (and today's life-world view of) separation of technology from nature—and, conversely, the separation of nature from technology. His critique of the Aristotelian dichotomy represents a major contribution to the foundation of the modern era and to its convictions. For Bacon, it is neither necessary nor possible to outwit or escape from nature (notably nature's resting and self-movement) by the actions of craftsmen and man-made technology. Rather, technology is what is *possible* according to nature whereas nature is described by the sociomorphism "law"—and not by phenomena and outer appearances that are present in our life-world. Nature is everything that is law-like. Hence, Bacon draws technology under the wide umbrella of nomological nature and, in this sense, "naturalizes technology." Technology is nothing but nature governed by mathematical laws; it is impossible for technology to not be in accordance with nature, as posited in Aristotle's work. Therefore, man—in order to construct technology—must be "servant and interpreter" of nature. He has to achieve knowledge about nature and to obey nature or, more precisely, nomological nature (NO I: 45/Aph. 1). In phrasing *natura* (enim non nisi) *parendo vincitur* (NO I: 46/Aph. 3), Bacon puts "obeyed" ("parendo") before "commanded" ("vincitur"): Nature to be commanded must be obeyed! However, the nature that Bacon wishes to subsequently "vanquish," "conquer," or "command" in order to reveal its secrets and shape it is different from the one he "serves" or "obeys." The nature to be vanquished by the means of technology is the type of nature that confronts man as an alien force in his day-to-day life: phenomenal nature, which includes, for example, the pest. In sum, Bacon (a) extends the general understanding of nature to encompass many branches of reality, in particular technology. In this sense, he is a precursor of a *naturalization of technology.* (b) On the other hand, nature has to be vanquished, commanded, or shaped by technological means. Thus, we also find in the work of Bacon the initial point of the modern path towards a broad *technologization of nature.*

Another aspect needs to be mentioned here. Bacon criticizes the Aristotelian *Four Causes* doctrine. Most prominently, he presents what was to become characteristic for modernity and the standard mindset in the 20th-century philosophy of science: a critique of the final cause, which "is [so] far from being beneficial that it actually corrupts the sciences" (NO II: Aph. 2). Finality, according to Bacon, is scientifically non-perceivable; this means that it is just metaphysics or, equivalently, fiction. After all, in nature

> nothing really exists [...] besides individual bodies, carrying out pure, individual acts according to law, yet [...] this very law, and the investigation, discovery and explanation of it, is the very foundation [...] of knowing as it is of operating.
>
> (NO II: Aph. 2)

So Bacon already introduces an understanding of laws into the core of the novel, experimental natural philosophy, or of modern science. The goal is to find "the general and fundamental laws which constitute forms" (NO II: Aph. 5). Admittedly, the notion of form may still sound somewhat Aristotelian; but Bacon reinterprets forms as underlying laws. To this end and entirely in accord with modern science, Bacon presupposes that a lawful "union [= unity] of nature" is the "foundation for the constitution of sciences" and, by this, "start[s] to set up the sciences" (NO II: Aph. 27).

The concept of nature advocated by Bacon can be seen in a reductionist light. According to Bacon, such a reductionism to basic properties is inherent in the structure of nature. Accordingly, "every natural action is carried out [by things infinitely small ...], by bodies too small to impinge on the sense" (NO II: Aph. 6). Yet Bacon does not defend an ontological atomism; he is more concerned with basic properties than with entities and is more focused on laws/forms than on atoms. However, none of his conceptions excludes in any way the self-activity and self-organization of nature. "Nature" is capable of doing and acting "from within," Bacon states (NO I: Aph. 4). This point is also stressed by Aristotle, who defines nature by its momentum and rest in and of itself. Surprisingly, Bacon assumes, in line with Aristotle, that there is self-creativity within nomologically conceptualized nature—a point that can be related to the recent discourse on technoscience (cp. Carrier 2011; Schmidt 2011b). Bacon acknowledges the emergence of new properties and qualitative aspects in the overall process of nature—concurring with Aristotle and with Schelling but surprisingly disagreeing with Newton. Bacon's line of argument comes very close to the present-day idea of the self-productivity of (naturalized) technology based on self-organization and a certain view of nature, namely that "nature is applied technology" (Carrier 2011, 53). That reflects a novel concept of "the technoscientific ontology of nature as a human creation" (ibid.).

In essence, Bacon claims that nature is nature insofar as it is governed by laws; nature in its core is nothing but mathematical law. Thus, techno-objects are not unnatural or do not go beyond nature; they are nature although they are constructed and created: Techno-objects are nature at work.

Technoscience as object-oriented interdisciplinarity

Besides the above-pursued historical perspective introducing Bacon's call for "fruit-bearing knowledge," which shows that an instrumental orientation is rather old, the recent vibrant discourse on "technoscience" might shed further insight into the ontology of techno-objects and of object-oriented interdisciplinarity; furthermore, the discourse on "technoscience" and Bacon's program of science have much in common. Although the notion of inter- or transdisciplinarity is hardly the focus of scholars working on "technoscience," the phenomenon itself is quite ubiquitous (Nordmann et al. 2011). Essentially, technoscience, on the one hand, and object-oriented interdisciplinarity, on the other hand, are to be considered twin sisters.

To start with, the very term "technoscience" is a composite. It signifies a strong interlacement of "science" and "technology" or of "natural science," "technical science," and "engineering." The present-day amalgamation makes it hard, or even impossible, to separate one from the other—that is the basic observation in the discourse surrounding technoscience.[15] In addition, there is more to it than simply the perception that, for example, many fields of physics and mechanical engineering are today inherently interwoven. More fundamentally, the point of departure of those who engage in the discourse on technosciences is—in line with what has been discussed so far under the label "interdisciplinarity"—the recognition that many of the recent, technically created objects as well as their design, creation, production, and use do not fit the mindset of our culture or of modern philosophy.

We seem to be experiencing a dissolution of (traditionally assumed und well-established) boundaries, such as the boundaries between natural and engineering sciences, between pure and applied research, between curiosity-driven and utility-oriented approaches, between the given and the fabricated, between theory and practice, and between science, technology, and society. Many scholars see the boundaries of the academic disciplines or those between the academy and society vanishing. In light of the transgression, and even elimination, of boundaries within the technoscientific regime, Paul Forman (2012) goes so far as to diagnose an "anti-disciplinary period" and an "end of modern sciences": "postmodernity entails anti-disciplinarity." Here we can identify a strong connection between the two discourses—namely the one on technosciences and the one on interdisciplinarity. In addition, technoscience and its object-oriented interdisciplinarity are interlaced with Baconian ambitions and the hubris of power: controlling, shaping, constructing, and creating the world.[16] Donna Haraway underscores that "technoscience is about worldly, materialized, signifying and significant power" (Haraway 2003, 43).

Such a view focusing on technical objects as real things in the networks of power is compelling. Although "technoscience" is often seen primarily as a diagnostic or reflexive notion, in that it is a term enabling reflection on the current state of science, technology, and society (Nordmann et al. 2011), it should ultimately be looked at from an ontological perspective, as Martin Carrier (2011) argues. A key observation made by Carrier and others is that the new technoscientific things elude, ontologically, a conceptual classification. Should the techno-objects, for example, be regarded as natural, as technical, or as cultural things? Should they be researched by the natural, engineering, or cultural sciences? Do these techno-objects enable us to fundamentally control the state of the world, or do they elude our control? Obviously, an irreducible "ontological indistinctiveness" remains (Daston and Galison 2007). In a strong sense, the ontological indistinctiveness holds, in particular, with regard to biotechnical objects. These technical systems appear to be and behave like nature according to the Aristotelian concept. They possess within themselves a principle of motion and change and of rest and cessation of change. Sometimes they grow like living organisms.

Well-established dichotomies that are still omnipresent in our life-world are now dissolving: Technology can be deemed bio-naturalized and culturalized; nature seems to be culturalized and technologized; culture is technologized and scientificized, and so on. In this line of thought, Haraway (1991) recognizes various entanglements. She gives the examples of "cyborgs" and "hybrids" such as test-tube embryos, trans-genetic organisms, in vitro fertilization, egg donation, artificial insemination, surrogacy, and human cloning. In a similar vein, Karin Knorr Cetina (1999, 149) uses the term "ontology" when she maintains that the expression of "molecular or cellular machines [...] can serve as a master analogy for the ontology of objects in the laboratory: the objects that stand out are not used as organisms, they are implemented as machines." Nicole Karafyllis (2007) coins the far-reaching fundamental term "biofact" to underline an ambivalent phenomenological indistinctiveness of organisms from technical systems in the field of bio-engineering practice. Following this observation, Christoph Hubig and Sebastian Harrach (2014, 41) argue that "on the basis of self-organization processes" a novel type of technical/natural hybrid that can be called "transclassic or informal" is emerging. It "blurs the interface between the technical, the natural and the human." Nature itself is seen to be productive, constructive, or creative. The German idealist Friedrich Schelling spoke about "nature as productivity"—taking up Baruch Spinoza's notion of *natura naturans*. From a technomorphic point of view, as I have already touched upon, nature is framed as "applied technology" (Carrier 2011, 53).

It seems undisputed among scholars that a hybrid ontology or ontology of hybrids needs to be developed—as a novel mindset in order to cope with "transclassic technology." As Hubig and Harrach (2014) also note, nano-techno-objects, in addition to the above-discussed biotechnological objects, again serve as excellent examples. Advanced nanostructures are capable of self-organization, self-growth, or self-assembly. In his famous book *Engines of Creation*, Eric Drexler (1990) presents a highly disputed futurological version of self-organization and bottom-up emergence. Nanobots, or molecular assemblers, are considered to be the constituents of "soft" or "molecular machinery," including "molecular fabrication" based on "autonomous self-organization."[17] The example of nanotechnology shows that the "ontological indistinctiveness" is not limited to bio-based technology. It reaches much further, even beyond nanotechno-objects, and encompasses, for instance, information and communications technologies. Take, for example, machine learning and artificial intelligence systems, in particular the "autonomous stock agents" common in high-frequency trading at the stock exchange.[18] Based on principles of self-organization, the algorithms are trained on real big data in the context of application; they learn in an open environment to make traders' "decisions." They are the abstract machines behind the scene, the "market makers" governing today's global neo-capitalist finance market. Such algorithmic techno-objects show human- or nature-like characteristics; in fact, their capacities come close to those of human actors.

The described ontological view on present-day technoscientific objects entails an epistemological concept having at least some realist or pragmatist grounds. At the same time, it clearly offends those concepts of traditional realism that presuppose the existence of human-independent nature and adopt a detached observer's perspective. The ontological view advanced in this book questions the dichotomy between constructivism and realism. "To be a construct does not mean to be unreal or made up; quite the opposite," Donna Haraway (2003, 46) argues. In a similar vein, Bruno Latour (1990, 71) maintains, "A little bit of constructivism takes you far away from realism; a complete constructivism brings you back to it." As we have seen, the epistemological position advanced by Latour (and in this chapter) is one that could be called real-constructivism. This position encompasses a commitment to a materialist or realist anchor. However, the implied epistemic truth is not related solely to sound explanations based on a reliable theory. The deductive-nomological model of explanation turns out to be far too narrow for the truth of the technosciences; mechanistic explanations that fulfil, in addition, the requirement of stability with regard to perturbations might be adequate concepts.

Epistemic truth in the realm of technoscience is engendered by the constructed new phenomena, built things, or produced objects themselves. In other words, truth emerges immediately from objects, things, and works: as the truth of the things and as the objectivity of the techno-objects—a point we have already seen in Bacon's works. To paraphrase Ludwig Wittgenstein, the world is all that is the case and what is the case includes the existence of a novel object. Alfred Nordmann (2006) discerns a "collapse of distance between representation and its objects." Davis Baird (2004) forges the term "thing knowledge," contrasting this with theoretical knowledge and its reference to deductive-nomological explanations. Truth is constructed and created with, by, and within the objects; it is represented *and* demonstrated by capabilities to intervene in and to change what is given. If you can construct an object, the knowledge underlying the action is to be considered true. According to Ian Hacking (1983), intervention, construction, and creation should be considered the central statement of the position of realism. Facts *and* artefacts are interlaced, so are reality *and* constructivity, truth *and* existence, knowledge *and* power. The well-established dichotomy between theory and practice is blurred—it probably never existed and was only a self-stylization. Nano-technoscience, for instance, manifested its truth in the molecular plot of the letters "IBM." In consequence, epistemic truth does not necessarily go hand in hand with new theoretical concepts of the nanocosm or with a sound body of deductive-nomological explanations. In this context, we could add that not only is the "thing knowledge" in technoscience created within and by technical systems, but the perception of the related phenomena also depends on apparatuses and instrumentation. "We observe objects or events with instruments. The things that are seen in twentieth-century science can seldom be observed by the unaided human senses" (Hacking 1983, 168).

What holds for science in general is even more obvious for technoscience: Without intervention, creation, and construction, a scientific methodology cannot be developed. Taking experimentation and intervention into account entails framing technoscience from an action-theoretical perspective: as a practice of action involving various actors or "actants" in different epistemic cultures (Latour 1993; Knorr Cetina 1999). Technoscience fabricates knowledge; it creates and constructs facts. The focus is the "activity of making [techno-] science and not the definition given by scientists or philosophers of what science consists of" (Latour 1987, 174). We therefore need to consider the "context of discovery" or, more specifically, the "context of construction and creation"—in contrast to the traditionally highly esteemed "context of justification" with respect to propositions, laws, theories, models, and concepts. The focus on the "context of construction and creation" has also been adopted by ethnographic researchers, who claim to have opened Pandora's black box of science or technoscience (Latour 1999). The more we look at the sciences or technosciences, the greater the heterogeneity becomes—revealing a (fruitful and productive) disunity (Galison and Stump 1996; Cartwright 1999). In essence, technoscientific action and truth seem much more complex and cannot be decontextualized. Bruno Latour (1987, 174) uses "the word 'technoscience' from now on, to describe all the elements tied to the scientific contents no matter how dirty, unexpected or foreign they seem."

In addition to the (techno-)object orientation just discussed, another point—characterizing the technosciences and their Baconian concept of research—merits consideration. The technosciences are guided by trans-epistemic values and, more specifically, in many cases by the economic demands of technological innovation in the global capitalist market. As such, the technosciences have a strongly instrumentalist perspective. In the technosciences, knowledge is judged by its utility with regard to the field of economics or business. Certainly, understanding knowledge in such a way opens the production of knowledge to commercialization—a perfect match with our neoliberal times. In the age of technoscience, science is in the marketplace (cp. Carrier 2011; Gethmann et al. 2015). Put differently, technosciences—or, similarly, mode-2 sciences—generate knowledge in "broader, transdisciplinary social and economic contexts" (Gibbons et al. 1994, 4). Clearly, technoscientific knowledge is not pure and value-free; it is not deemed an end-in-itself but a means—and a medium for continuous innovation. It is a highly acknowledged instrument for obtaining competitive advantages in the global capitalist market and for ensuring growth and wealth: Technoscientific knowledge is tantamount to the power to change the world; it has a transforming impact on our societal life. In sum, the relevance of technosciences is apparent from the dominance of technoscientific knowledge in present-day societies in the neoliberal age; our societies are essentially knowledge societies.

The above-discussed list of points characterizing technoscience is not exhaustive. However, it will suffice to substantiate our observation of a common core shared by the technosciences themselves, Bacon's concept of sciences as research, and techno-object–oriented interdisciplinarity. The overarching characteristic is that well-established categories and presupposed dichotomies became blurred once science went "to the marketplace." Whether we can, and whether we should, re-establish the categories is a matter at issue. Ulrich Beck and Christoph Lau (2004), for example, see a societal need for "boundary-setting politics" in order to regain orientation: In the technoscientific age, a reflexive modernization, including a *good societal life*, seems simply impossible without boundaries. What Beck and Lau have in mind is a shift from object-orientation towards a more reflexive and critical attitude that is part of problem-oriented interdisciplinarity. Such a viewpoint will be discussed in detail later on in this book.

Summary and prospect

Bacon's influence on present-day research—notably on the regime of technosciences—can hardly be overestimated.[19] More than two decades ago, the physicist Michio Kaku (1998, 16f), in line with Bacon's optimistic view, envisioned with a hubris-like air of elevation that

> for most of human history, we could only watch, like bystanders, the beautiful dance of Nature. But today, we are on the cusp of an epoch-making transition, from being passive observers of Nature to being active choreographers of Nature. [...] The *Age of Discovery* in science is coming to a close, opening up an *Age of Mastery*.

Although such enormous capabilities of humans to control and master the things are still futuristic dreams, this certainly seems to be the direction in which technoscientific research is heading. Technoscience is research guided by trans-epistemic values. It centres on the creation, construction, design, and production of interdisciplinary techno-objects. In sum, object-oriented interdisciplinarity can be seen as an ambivalent pacemaker putting the Baconian program fully into practice.

Bacon's instrumentalist view of science and his materialist real-constructivist epistemology are now, in essence, more powerful than ever before, especially in the growing field of interdisciplinarity and interdisciplinary technosciences. Framed from this angle, techno-object–oriented interdisciplinarity has a rather long history; it is by no means novel. "Clearly," the Dutch philosopher Hans Achterhuis (2001, 2) points out, "Bacon's observation about the transforming impact of technology, made at the beginning of the seventeenth century, is as topical as ever." Yet, although Bacon's program is anything but new and dates back to the 17th century, we are only now at the stage of witnessing the fulfilment and implementation of the Baconian goals (Muntersbjorn 2002).

To summarize more specifically, central ideas of today's discourse about technoscience and about (object-oriented) interdisciplinarity can be traced back to Bacon. The *motive* behind and the *approach* of object-oriented interdisciplinarity link light-bearing with fruit-bearing knowledge, the end-in-itself with the means-to-an-end, and curiosity-driven insight with utility-oriented application. Experimenting, manipulating, constructing, and creating are central to the real-constructivist's *method*; experimenting mediates between the given and the constructed as well as between object *and* subject, mechanism *and* action. *Manifestation, evidence,* and *truth* of knowledge emerge through works and machinery, as the truth of works. In the latter, facts *and* artefacts, reality *and* constructivity, theory *and* technology, and also truth *and* utility, knowledge *and* power are combined. In other words, techno-objects and not theories are in focus.

The *Philosophy of Interdisciplinarity* introduced in this book endeavours to disclose the historical origins of object-oriented interdisciplinarity. Consideration and acknowledgement of the Baconian background behind this central stream of interdisciplinary research, as we see it today, can be understood as the basis for a broader awareness of its ambivalence and for a more profound critique. Basically, we have to go through Bacon and deal with his program—in order to go beyond him. Such a realization is central to a critical-materialist stance as that taken in this book: "A critique of society requires a critique of knowledge and vice versa" (Adorno 1969, 158).[20]

Notes

1 In the early 1970s, Jantsch (1972, 103) stressed: "Above all, we must ask: What is the purpose, interdisciplinarity?"
2 For the discussion on "ontopolitics," see Chandler (2018).
3 Scholars from very different schools critique this reduction, see, for example, Dewey (1929), Horkheimer and Adorno (1972), Hacking (1983), Latour (1987), Rouse (1987), Ihde (1991), Janich (1992), Baird (2004), Nordmann (2008a), Cozzens and Gieryn (1990), Haraway (1991), Feenberg (1991, 2002), Radder (2003), Ihde and Selinger (2003), Daston and Galison (2007), and Zittel et al. (2008).
4 If there is any historical epochal break, as many scholars recognize, then it is by no means a rupture in the programmatic thoughts or ideals but in the realization, fulfilment, and accomplishment of the program today with regard to novel kinds of techno-objects: If anything, we are witnessing an epochal break with regard to the "ontology of the objects." This observation concurs with Martin Carrier's perception that "the most important reorientation of recent science does not arise at the methodological level but at the ontological one" (Carrier 2011, 51). In other words, the "reorientation of science" takes place with regard to techno-object–oriented interdisciplinarity.
5 This statement holds, although present-day interdisciplinary objects might elude control (cp. Kastenhofer and Schmidt 2011).
6 My translation of the original leitmotif of the Frankfurt School (J.C.S.).
7 The German term "Revolution der Denkungsart" is sometimes translated as "intellectual revolution."
8 Kant writes in German: "Heeresweg der Wissenschaft." A closer translation of "Heeresweg" would be "way of the army" rather than "highway."
9 What Bacon proposes is therefore not "easy to [...] explain": "people will still make sense of things new in themselves in terms which are old" (NO I: Aph. 34). The abbreviation "NO I" used throughout this chapter refers to Novum Organon, Part I; "NO II" refers to Novum Organon, Part II. I use the translation by Graham Rees and Maria Wakely: Bacon (2004) (1620): *The Instauratio magna Part II: Novum organon and Associated Texts*; (eds. Graham Rees and Maria Wakely) Clarendon Press, New York. I do not follow the translation of Lisa Jardine and Michael Silverthorne in Bacon, F., 2000 (1620): *The New Organon*; (ed. by Lisa Jardine and Michael Silverthorne) Cambridge. The classic translation is by James Spedding, Robert Leslie Ellis, and Douglas Denon Heat: Bacon, F., 1863 (1620): *Novum Organon* (standard translation by James Spedding, Robert Leslie Ellis, and Douglas Denon Heath); in The Works (Vol. VIII), Taggard & Thompson. A newer edition is by Fulton H. Anderson: Bacon, F., 1960 (1620): *The New Organon and Related Writings*; (ed. Fulton H. Anderson); Howards W. Sams & Co., Indianapolis.
10 We have shown that this kind of transdisciplinarity is not synonymous with what is called problem-oriented interdisciplinarity, see Chapter 2 of this book.
11 Implicitly, Bacon alludes to disciplinarity as a way to facilitate a certain, namely object-oriented type of interdisciplinarity. Two requirements (boundary setting *and* boundary crossing) are necessary for a semantically adequate use of the notion of "interdisciplinarity."
12 Bacon's and Dewey's critiques were, in principle, supported in the early 20th century by advocates of the Copenhagen interpretation of quantum physics (Bohr, Heisenberg, and others).

13 For more details, see Chapter 7.
14 Although Bacon is an empiricist, he shares many ideas with the critical-rationalist thinker Karl Popper.
15 Besides Hottois (1984), also Latour (1987), Haraway (1991, 2003), Ihde (2002), Weber (2003), and Nordmann (2005, 2008a) forged the term "technoscience." Nordmann's list of criteria defining "technosciences" fits very well to Bacon's program (Nordmann 2005).
16 This holds, even if, in one case or another, it might turn out that some techno-objects elude control (Kastenhofer and Schmidt 2011; Schmidt 2015b; Schmidt 2016).
17 According to Drexler (1990), the self-productivity of nano-assembly could serve as the basis for an engineering revolution in the fabrication of complex systems—leading to a new renaissance and a next industrial revolution of soft machines. "Assemblers will be able to make anything from common materials without labour, replacing smoking factories with systems as clean as forests" (Drexler 1990): Tiny gears, motors, levers, casings, and proteins, genomes, mitochondria, cells, organs will be produced by molecular tools in processes of self-organization.
18 Autonomous stock agents function on a mathematical-numerical basis consisting of artificial neuronal networks, genetic algorithms, or similar numerical structures of universal learning machines.
19 See Krohn (1987), Böhme (1993), Carrier (2001), Muntersbjorn (2002), and Carrier (2011).
20 My translation (J.C.S.). In the same vein, science and technology studies (STS) scholars and critical theorists also underscore that facts and artefacts are political ("artifacts have politics," Winner 1980); scientific knowledge and theories cannot be separated from society and societal institutions. According to this viewpoint, epistemology is also part of the power discourse. In classic STS terms: "Truth speaks to power" and "Power speaks to truth" (Collingridge and Reeve 1986; Jasanoff et al. 1994).

5 Society and societal problems
Conceptualizing problem-oriented inter- and transdisciplinarity

Addressing real-world problems

Inter- or transdisciplinarity is often deemed to offer answers or even solutions. Considering that answers or solutions are given in response to problems, some interdisciplinarians define interdisciplinarity exclusively in reference to problems: "Any research field or research project that addresses real-world problems is considered to be essentially interdisciplinary" (Krohn 2010, 33).[1] Whereas other scholars do not go as far as that, many—particularly those who advocate transdisciplinarity[2]—make reference to pressing societal and ethical problems in one way or another.

Because problems are often supposed to be simply given and to be solvable, the scholarly parlance relating to problems might suggest that the problem-oriented type of interdisciplinarity can be exclusively framed using a means–ends schema; it seems instrumentalist.[3] However, that gives an incomplete first impression. Problem-oriented interdisciplinarity can, but need not always, be means-centred. It is not to be defined solely by its orientation towards providing a definite or ultimate solution to a societal or ethical problem—even if that is what some problem-oriented projects seek to accomplish. In particular, for the so-called "wicked problems," there is no solution or at least no stopping rule to determine when the problem has been solved (Rittel and Webber 1973).

Although many problem-oriented projects are clearly instrumentalist—as such, they accept given problems and aim to deliver solutions—problem-oriented interdisciplinarity is a much richer and more multifaceted type of interdisciplinarity, as will be shown. It offers far-reaching perspectives and can pursue more than an instrumentalist approach: critical reflection on the setting, identification, framing, and consideration of problems—which represents an essential step towards critical-reflexive interdisciplinarity (as a subset of the problem-oriented type). Critical reflection includes reflection on the points of departure and agendas of research projects; on the institutionalization of academic practice; on the criteria for determining what can be legitimized as scientific knowledge; on the claims of truth, expertise, and authority of the science system; and, of course, on the role and accountability of the academy in society.

DOI: 10.4324/9781315387109-5

In general, in problem-oriented interdisciplinary projects, problems are given or constituted within a broader societal context, which includes a variety of communities and sometimes (but not always) also the public. For some scientists, such an exposure to non-epistemic values of heterogeneous communities might turn out to be a double-edged sword. Their expertise, disciplinary identity, and even entire discipline could be at stake. To meet this challenge—and this is a crucial step for transgressing an instrumentalist approach—critical self-reflection on one's own approach to a problem is required. Such self-reflection includes a consideration and awareness of one's own disciplinary procedures of knowledge production as well as of the blind-spots, boundaries, and limits of one's own disciplinary approach. That is not to say that problem-oriented interdisciplinarity always encompasses this critical reflexivity. However, it does provide an opportunity to engage in reflection at a deeper level. To illustrate this crucial point, we need to carry out a more differentiated analysis of problem-oriented interdisciplinarity—an analysis that also goes beyond the question of whether and to what extent lay people or societal stakeholders should be involved in the process of knowledge production.

The problem with "problems"

The buzzword "problem" is not very specific. Problems seem omnipresent. Karl Popper (2000, 97) maintains that "life is always problem solving." Referring to the research and science system, he adds, "we study not disciplines, but problems. Often, problems transcend the boundaries of a particular discipline."[4] However, academic problems are typically identified, framed, worked on, and solved in the disciplines or within disciplinary frameworks or paradigms. For example, physics, biology, medicine, engineering, and sociology all appear to be in some way problem-centred or problem-driven. But even in this limited disciplinary context—and not solely across the wider horizon of interdisciplinarity—the notion of "problem" itself remains vague.

Notwithstanding this vagueness, "problems" often show up as a central element in the definition of problem-oriented interdisciplinarity or, in a more encompassing sense, transdisciplinarity.[5] Julie Thompson Klein et al. (2001), for example, characterize transdisciplinarity in reference to societal problems: transdisciplinarity is "joint problem solving among science, technology, and society." In the same vein, Jürgen Mittelstraß (1998, 44) underlines that "by transdisciplinarity we describe types of research and sciences that transcend disciplinary orientation in a problem-oriented mindset."[6] Jochen Jaeger and Martin Scheringer (1998) argue that interdisciplinarity should be understood as a "problem-related form of science" and as a "problem orientation without method constraints." Egon Becker and Thomas Jahn (2006, 310) conceptualize social ecology as a "transdisciplinary research program" in which "challenging problems are not simply given via the disciplines, but are deeply rooted in societal practice."

Similarly, Gotthard Bechmann and Günter Frederichs (1996) regard inter-disciplinarity as "problem-oriented research in between science and public policy" and focus in particular on interdisciplinarity in the domain of technology assessment and scientific policy advice.[7] According to Michael Decker (2010, 145), methods of technology assessment are essentially inter-disciplinary because they "identify and work on trans-scientific problems" that are "political or societal problems." Günter Ropohl (2005, 29) argues along similar lines that "transdisciplinary sciences define their problems with regard to life-world relevance."[8] Armin Grunwald (2004) seeks to identify quality criteria to specify the added value of "issue-driven interdisciplinary and problem-oriented research." Obviously, problems are ubiquitous.

In scholarly use, however, the term "problem" is vaguely defined; it remains a buzzword lacking clarity. The objective of this chapter is to facil-itate the conceptual clarification of problem-oriented interdisciplinarity (and equivalently of problem-oriented transdisciplinarity); the chapter also reviews the typical mainstream understanding of this kind of interdiscipli-narity (i.e., the shortcoming of the instrumentalist perspective with its focus on means, methods, and recipes). By advancing a critique, I wish to support the prospects of *critical reflexivity* in the debate on interdisciplinarity.

Wicked problems

It is often intoned among interdisciplinarians that "the world has problems," particularly societally "complex" or "wicked problems," but the academy has disciplines and the university has departments: The problems of the real world and those of the academy are incommensurable. The perception of incommensurability serves as the point of departure for the advocates of problem-oriented interdisciplinarity.[9]

In order to explain why interdisciplinarity is deemed necessary, challeng-ing complex, wicked, real-world problems are at focus. This rhetoric has its own history. Participants in the Manhattan Project, for instance, did not use the term "interdisciplinarity," although the project can be classed as prob-lem-oriented; its original intention was to counteract the development of a nuclear bomb by Nazi Germany; a possible victory of Nazi Germany was regarded as the problem (Bush 1945; de Solla Price 1963). In subsequent approaches, Alvin Weinberg (1972) and also Horst Rittel and Melvin Web-ber (1973) were among the first to suggest the term "problem" to describe a specific orientation of science *in* and research *for* society.[10] Weinberg (1972) speaks of "issues" and "problems" like the challenging and pressing questions raised by "the deleterious side effects of technology" (e.g., the safety of nuclear power plants). Weinberg's still-relevant diagnosis is that there are central questions "which cannot be answered by science" with its traditional intra-academic orientation. He proposes "the term 'trans-scien-tific' for these questions since [...] they transcend science." In developing his concept of "trans-science" (ibid., 209), which shares much with today's notion of "transdisciplinarity," he argues that "scientists have no monopoly on wisdom where this kind of trans-science is involved: they will have to accommodate to the will of the public and its representatives" (ibid., 222).

Weinberg's use of the notion of "problem" is more or less implicit. Rittel and Webber (1973), on the other hand, explicitly harness this notion and coin the term "wicked problem":[11]

> We use the term 'wicked' in a meaning akin to that of 'malignant' (in contrast to 'benign') or 'vicious' (like a circle) or 'tricky' (like a leprechaun) or 'aggressive' (like a lion, in contrast to the docility of a lamb).

They characterize a "wicked problem" using the following criteria (ibid., 161 ff.): (1) There is no definitive formulation of a wicked problem; (2) wicked problems have no stopping rule; (3) solutions to wicked problems are not true or false; rather, they are good or bad; (4) there is neither an immediate nor an ultimate test of a solution to a wicked problem; (5) every solution to a wicked problem is a 'one-shot operation' because there is no opportunity to learn by trial-and-error and every attempt counts significantly; (6) wicked problems do not have an enumerable (or an exhaustively describable) set of potential solutions, nor is there a well-described set of permissible operations that may be incorporated into the plan; (7) every wicked problem is essentially unique; (8) every wicked problem can be considered a symptom of another problem; (9) the existence of a wicked problem can be explained in numerous ways, and the choice of explanation determines the nature of the problem's resolution; and (10) the planners and actors tackling the wicked problem have no right to be wrong (cp. Rittel and Webber 1973, 161 ff.; see also Norton 2005). As Rittel and Webber underline, this list is not exhaustive. However, these characteristics distinguish "wicked problems" from intra-academic disciplinary ones.[12]

From a cognate perspective, we may recall Erich Jantsch's (1972) suggestion of a "purposive understanding of interdisciplinarity" and "purpose-oriented interdisciplinarity." Jantsch refers to "systemic problems" and argues against the "danger" that can "be seen here in the temptation to take a straightforward (non-systemic) problem solving approach" (ibid., 113/99): "There are no clear-cut problems to be solved. The classical single-track and sequential problem-solving approach itself becomes meaningless today. This may come as a 'cultural shock' to our pragmatic and efficient society, valuing nothing more highly than 'know-how'" (ibid., 101). A recipe or a clear method does not exist. In essence, interdisciplinarity does "not [stand] for problem-solving, but for a continuous process of profound self-renewal" (ibid., 102). Here, Jantsch digs deeper than Weinberg. He sees explicit reflection on, and the review and possible revision of, purposes as the highest level of interdisciplinarity, which he calls transdisciplinarity.[13]

As shown, the problems addressed by problem-oriented interdisciplinarity are deemed external to the disciplines or to the academy. They are considered societal or ethical problems that are pre-defined by society (e.g., lay people, politicians, or other stakeholders). Today's science-based, future-centred approaches, such as technology assessment, sustainability science, and global change research, likewise take a societally relevant point of departure for research endeavours. Because problems precede both the *context of discovery* and the *context of justification* (of knowledge), problem-oriented interdisciplinarity is a specific type that cannot be subsumed under the label of method-, theory-, or object-oriented interdisciplinarity. A strong intentional or teleological structure of the process of knowledge production is assumed in problem-oriented interdisciplinarity. The very first step in scientific inquiry—identifying a problem and setting an agenda, interlaced with the intention to obtain knowledge—is often (mistakenly) seen as a contingent factor. It has been widely ignored or devaluated by philosophers, in particular by philosophers of science. There are exceptions. The school of discourse ethics, developed by Karl-Otto Apel and Jürgen Habermas, has not followed this mainstream of neglecting the very starting points. Habermas (1993), for example, focuses on the purposes and interests driving research activities. In addition, concepts such as rationalist technology assessment have addressed the issue of problems as the starting point of any problem-oriented interdisciplinary project (Grunwald 1999; Gethmann et al. 2015; Lingner 2015).

These approaches represent exceptions. So what is behind the common disregard for the initial stage and the point of departure of research activities? It seems to be a natural consequence of the predominance of analytical philosophy of science as well as of the predominant public image of the sciences. Analytical philosophers of science have always been reluctant to consider trans-epistemic aspects within science's core and its claim of knowledge and truth. They more or less parade a traditional, value-free ideal of the sciences. But as soon as the notion of problem is included, it becomes clear that normativity is in some way involved. Consequently, framing knowledge production from the angle of problems contributes to a critique of the self-stylization of science as a value-free enterprise. Robert Frodeman (2014, 44) is even more provocative in maintaining that the "problem definition is often an irredeemably political and axiological matter where there is no neutral answer to be had." The notion of problem therefore can be regarded as a point of entry to critical reflection that begs an explication of *who* is defining *what* as a problem and *why* or *for what*. In fact, problems constitute a kind of normative–descriptive hybrid carrying a call to action, since problems—in science and society alike—are seen in a negative light, indicative of a deficit state needing to be tackled.

Although philosophers have not addressed the notion of problem explicitly, the word occurs in the works of many famous philosophers. Karl Popper stresses that a good hypothesis has to include "risky problems," Thomas S. Kuhn believes a "paradigm determines the choice of problems," Imre Lakatos coins the notion of "progressive vs. degenerating problem shift," and Larry Laudan views "sciences as problem-solving action." Popper (2000, 97) also points out that "we aren't studying things, but problems."[14] However, like other philosophers of science, he does not provide a definition of the term "problem." What is most interesting is that Kuhn presents an idea that comes close to the recent discourse on problem-oriented interdisciplinarity. Even back in 1962, Kuhn perceived a professional blindness among scientists in regard to societal problems: "A paradigm can [...] even insulate the community from those socially important problems that are not reducible to the puzzle form, because they cannot be stated in terms of the conceptual and instrumental tools the paradigm supplies" (Kuhn 2012, 37). Furthermore, he adds, "the really pressing problems, e.g., a cure for cancer or the design of a lasting peace, are often not puzzles at all" (ibid.).

In Kuhn's terminology, problems should not be regarded as puzzles since they do not have clear and ultimate solutions in the way that scientific puzzles are assumed to have. The distinction between problems on the one hand and puzzles and their solutions on the other hand is relevant to the discourse on interdisciplinarity. Problem-oriented interdisciplinarity does not always offer solutions. It is rather the case that, when viewed from a critical-reflexive angle, much is achieved if and when a problem is identified, framed, constituted, clarified, and addressed. Problem-oriented interdisciplinarity can deliver advice on how to tackle and to handle a problem, but it does not always solve the problem itself. It supports but does not actually provide decisions. In addition, the sciences, for their part, and also interdisciplinary projects are not legitimized to recommend and implement solutions. If that were the case, democratic societies would turn into expertocracies or technocracies.

Proposing an analytic clarification

To what does the term "problem" actually refer? The philosopher Gereon Wolters (2004, 347) defines a problem as the "incompatibility of some propositions (the 'problems') with the set of those propositions that are considered as true or evident."[15] In other words, a problem is what does not fit into the body of accepted knowledge; it challenges a certain claim of knowledge. The notion of problem therefore can be conceived of as a concept of relations: It is based on the relation between two or more propositions.

This approach from the angle of philosophy of science, including the philosophy of language, is a necessary but insufficient attempt to clarify what a problem is. The definition should not be restricted to propositions or to general cognitive aspects. The philosophy of science needs to be complemented with philosophic action theory and ethics in order to give further substance to the notion of problem. Social scientists, such as Dietrich Dörner (1995), Christian Pohl together with the philosopher Gertrude Hirsch Hadorn (2007), Roland W. Scholz (2011), and Uwe Schneidewind and Mandy Singer-Brodowski (2013) have, from different angles, developed such integrative approaches. Although Dörner does not focus explicitly on interdisciplinarity, his concept of problems can serve as a framework for the clarification of what comprises problem-oriented interdisciplinarity. According to Dörner (1995), the notion of problem can be regarded as a relation of three elements which encompass normative and descriptive dimensions: an undesired initial state or current situation, including an anticipation of prospective futures; a desired final or ultimate state describing what the future should look like; and a barrier, obstacle, or hurdle that prevents or inhibits the transformation of the present state into the future state.

Christian Pohl and Gertrude Hirsch Hadorn (2007) and Roland Scholz (2011) take a similar stance, but they go further than Dörner by assigning a pivotal role to each piece of knowledge. Although their notion of problem is not elaborated analytically in detail, we can utilize their three types of knowledge to define a problem as follows: A problem exists if, and only if, there is an appreciable difference or discrepancy between

1. the *system knowledge*, which reflects or represents (in a more descriptive approach) the whole object or system under consideration including the current state and knowledge-based scenarios for the evolution of the current state (without taking action), on the one hand ("actual situation"), and
2. the *target knowledge*, which refers (normatively) to the desired state in the future of the system ("target state"), on the other hand,

and, furthermore, if

3. we currently have no *transformation or action knowledge* that could enable the transition from the actual situation (ad 1) to the target state (ad 2)—in other words, action-enabling knowledge about how to overcome barriers and hurdles (cf. Pohl and Hirsch Hadorn 2007).

From the above, it follows that the definition of the notion of problem comprises at least three knowledge types—and has to be considered as being at least three-relational. It includes system knowledge and target knowledge and a non-existence of transformation knowledge. Framed from this angle, problems are clearly descriptive–normative hybrids.

In addition, a number of formal dimensions deserve to be considered when specifying the notion of problem. A temporal dimension should be included. As has been exemplified, problem-oriented interdisciplinarity contributes to the perception and framing of a situation or a phenomenon as a problem. The word "situation" can refer to an actual or current state or to an anticipated future state. Certainly, some future states may be largely undesirable—in the sense of a catastrophe or a dystopia—and the actual state may be the desired one (e.g., when we think of global change effects). Furthermore, in the case of *potentially* negative effects, a problem has not yet arisen, but it might emerge in the future; an anticipated potential, but as yet non-existent, problem is viewed as being 'real' today; it induces a call for action. This means that effective and relevant kinds of problem-oriented interdisciplinarity are those which are inherently future-oriented. They can be regarded as anticipatory, prospective, or precautionary research *ex ante*: The goal is to prevent problems from occurring, such as by means of a problem radar, based on a precautionary principle and supported by methods of prospective technology assessment (Liebert and Schmidt 2010).

In summary, problem-oriented interdisciplinarity provides system and target knowledge—and, above all, (positive) transformation knowledge, namely knowledge about how to enable the transformation and overcome the barriers or hurdles. It includes a temporal dimension and *ex ante* reflection on prospective future states. Bringing all aspects together, I speak of *problem* (or problem-oriented) *knowledge*. Although the balance and interplay among the three kinds of knowledge may remain debatable, it is undisputable that problem knowledge is a kind of action knowledge: Problems call for action. If one does not have what one wishes to have and is unable to procure it, one has a problem. If we desire to live in a world without atomic weapons or to travel without carbon dioxide emissions but are unable to do so, we are faced with problems. This notion of problem carries with it central elements of certain action theories, including aspects of "inhibited effecting" (Wright 1991) and the "thwarted realization of objectives and purposes" (Grunwald 2002). Such a concept of problem touches on various fields such as philosophical ethics, philosophy of science, decision and planning theory, technology assessment, risk research, and scenario methodology.

The implicit assumption: internal–external dichotomy

Let me now seek to provide additional clarification. In the discourse on problem-oriented interdisciplinarity, a conspicuous assumption regarding the existence of boundaries—namely of a demarcation between the sciences and society—is evident.[16] An internal–external dichotomy is presupposed; intra-scientific and extra-scientific domains seem to be separated by boundaries.[17] With this type of interdisciplinarity, the attempt to transcend the (assumed) boundary takes place in two directions: (*extra-scientific*) societal problems are identified, re-formulated, and transferred to the intra-academic domain (potentially in collaboration with public stakeholders and lay people) and, further down the road, the findings are transferred back to the societal domain in order to contribute to *extra-* or *trans*-scientific, societal problem solving or prevention. In the 1970s, the underlying thesis of an internal–external dichotomy was included in the "finalization thesis" advocated by Gernot Böhme et al. (1983): In certain phases during the evolution of the sciences, external goals are necessary to initiate and drive science's (internal) development. To employ Kuhn's terminology, external goals determine the post-paradigmatic (and also the pre-paradigmatic) phase of research activities. Similar dichotomies are found in other concepts today, such as in the fields of post-normal, post-academic, mode-2 or technoscience, or transformative research.

The kinds of problems addressed by problem-oriented interdisciplinarity are therefore not those that are internal to the academy—that is, not problems based on major interdisciplinary objects or things or problems existing solely at the interface between different scientific disciplines.[18] For these latter kinds of problems, the borderline between science and society is not so relevant. Conversely, the assumption that such a borderline or demarcation exists is indispensable within problem-oriented interdisciplinarity since this boundary must—by definition—be transcended. That is why problem-oriented interdisciplinarity sometimes may be considered transdisciplinary. It can also be viewed as a kind of science of translation—from extra to intra and vice versa. The problem has to be translated in a manner to constitute a scientific object: According to John Dewey (1929), the constitution or construction of the object of inquiry poses a major challenge to the sciences.

> The risky character that pervades a situation as a whole [namely the problem] is translated into an object of inquiry that locates what the trouble is, and hence facilitates projection [and development] of methods and means of dealing with it.
>
> (Dewey 1929, 223)

Unfortunately, this fundamental "translation"—the translation of the problem or constitution of the object of inquiry—has been neglected from a philosophical perspective.[19] In particular, philosophers of language have remained silent on the subject; criteria for a successful transfer or translation from one domain to another have not yet been developed. It is paramount to look at what takes place in the series of steps from (a) the perception of a societal problem, (b) the constitution of an extra-academic problem requiring scientific inquiry, (c) the re-formulation and translation into a scientific problem, and (d) the latter's decomposition into a discipline-oriented problem to (e) the synthetic procedures of back-translation into the extra-scientific, societal realm. Certainly, this is not a mono-causal process but rather an iterative one containing various feedback loops. Egon Becker and Thomas Jahn (2006, 310) employ the term "problem dynamics" to describe the multifaceted translation procedure between science and society as well as that within the sciences (cp. Hummel et al. 2017). Problem dynamics, in fact, proves to be the central issue in the methodology of transdisciplinary. According to Christian Pohl and Gertrude Hirsch Hadorn (2006, 40), "the core element of transdisciplinary research is the question of how problems are to be identified, framed and structured within a broad area under consideration."[20]

From the above discussion, it is obvious that problem-oriented interdisciplinarity has to face a number of methodological challenges—from the framing of the problem at the very beginning, to the various transfer and translation procedures, to the final outcome and its re-transfer to the extra-academic life-world. Compared with the norms applicable in the (well-established scientific) disciplines and governing intra-academic knowledge production, this type of "transdisciplinary knowledge" is expected to meet higher quality standards because of the requirements of both society and (various) scientific communities.

Epistemological positions

In addition to the formal issues pertaining to problems, the epistemological status of the latter is not immediately clear either. To obtain a more profound understanding of the philosophic nature of the debate surrounding problem-oriented interdisciplinarity, two contrary positions on the status of a problem—based on different epistemological viewpoints or intellectual schools—deserve to be taken into consideration.

Although problems are induced by human behaviour and are symptomatic for the Anthropocene, realists presuppose that problems exist in the "ontology" of the world, regardless of whether or not humans perceive or know about them, e.g., global warming or the loss of biodiversity.[21] Constructivists, by contrast, assume that problems do not exist outside of human perception. The constitution of something *as a problem* therefore becomes a matter of central importance. These two epistemological positions are also well-established in a similar debate, namely the debate on risks: The sociologist Ulrich Beck defends risk-realism, whereas Niklas Luhmann supports risk-constructivism. Each of the positions is fuelled by a different accusation directed at the other—that of "alarmism" versus that of "relativism."

Since both positions are—more or less—supported by sound arguments, I would like to propose here a middle ground that comes close to a pragmatist or even a critical real-constructivist approach. (a) A realist dimension seems indispensable. The reason for starting any problem-oriented interdisciplinary project is provided by something that actually exists. A real thing, certain subject matter, or specific fact constitutes the core of a problem. The hole in the ozone layer, for instance, is not a social or cognitive construct; it is an undesirable present state that undeniably exists. Global warming is not a figment of the imagination, it is a scientific fact.[22] A *minimal realist* position—that is to say, a *minimal realism*—underscores that problems are referential: Problems refer to a real-world situation. In particular, scientific knowledge—generated, for instance, in the field of earth, atmospheric, or environmental science—contributes to justifying that something needs to be seen as a problem. (b) However, the reference or connection to a really existing something is a necessary albeit insufficient condition for characterizing something as a problem. For example, nuclear power plants are real systems but their mere existence is not sufficient to suggest that a problem exists; they are not per se a problem. In this regard, as already outlined above, three kinds of *constructions* are indispensable for framing a situation or an object, namely the nuclear power plant, *as a problem*: the system construction, the target construction, and the transformation-barrier construction. All three prove to be cognitive constructions. (a) The *system construction* encompasses decisions on demarcation—what comprises the system and what comprises its environment. For instance, should we (and, if so, to what extent) regard human health and the state of the environment as part of the system we call "nuclear power plant"? Should we include non-civil (dual) uses and the proliferation of nuclear materials in our system? (b) The *target construction* refers to the goal-setting procedures and to the desired future state. (c) The *transformation-barrier construction* involves framing and analysing the barriers and obstacles hindering us from reaching the desired future state.

Normativity, ascriptions, and decisions play a crucial role in all three constructions. They can also be related to the three types of knowledge discussed in a preceding sub-chapter: system, target, and transformation knowledge.

It thus becomes clear that realism and constructivism come together in problem-oriented interdisciplinarity. I will tentatively call this pragmatist-centred epistemological position real-constructivism. Problems are ambivalent since they are *material* real-constructions—induced by manifest human activity in the Anthropocene—and, second, they are *cognitive* real-constructions—identified by those who perceive and aim to deal with them. Problems are constituted or constructed on the basis of *real* situations or matters of fact and according to normative criteria and goals. They thus can be viewed from an action-oriented perspective, as John Dewey (1929, 223) maintains: "the first and most obvious effect of this change in the quality of [certain scientific] action is that the dubious or problematic situation becomes a problem." In that sense, problems have a twofold aspect: They do exist (realism) *and* they are generated and perceived (constructivism) at the science–society interface.

Such an epistemological position clearly levels criticism at any kind of naïve problem realism. In the latter case, problems would be unquestioningly accepted as plain facts and perceived simply as tasks to be tackled. The presentation of case studies in some engineering ethics textbooks (e.g., Harris et al. 2005) might serve as an example of naïve problem realism. In a more extreme example, the question of whether execution by the electric chair is preferable to execution by lethal injection is formulated as a problem. Some of these textbooks do not address the underlying challenges involved in the ethical justification of the death penalty and execution as carried out by democratic states. Contrary to this way of thinking, we have seen that problems are not simply given. Problem-oriented interdisciplinarity rejects this kind of naïve problem realism. The central question to be answered at the beginning and during the course of every problem-oriented research project is: What is the precise nature of the problem at hand? The constitution of the problem should be negotiable by employing rational arguments with explicit reflection on and possibly revision of the normative criteria used to decide the problem's relevance—while never losing sight of the associated goals and purposes.

In light of these epistemological considerations and the methodological challenges that arise, we experience *problems with problems* on at least five levels in inter- or transdisciplinary projects; as Bryan Norton (2005, 136) puts it, we encounter "problems with the problem formulation." The *first level* involves an awareness of and sensibility towards the life-world. Before a phenomenon, a thing, or an event can be termed a problem, its actual existence has to be perceived or acknowledged. In this act of recognition, intuition and emotions play a central role—in other words, the first problem of problems is the perception and justification of the existence of a certain problem. The *second level* concerns explicitly the construction (constructivism) or framing (realism) of the thing or phenomenon *as* a problem. Decisions are required to be made on the basis of argumentation and justification, which includes explicit reflection on the presupposed normative reference frame. At this stage, problems have the character of normative–descriptive hybrids.[23] On a *third level*, a scientific object specific to the problem has to be constituted or created; put differently, the object needs to be demarcated from its environment. This entails translating the problem into a scientific question or issue that can be addressed by the sciences and by the scientific disciplines in particular. The *fourth level* concerns the agenda-setting and design of research processes. It includes the decision about who is in and who is out or, more specifically, whose scientific discipline, knowledge, and competencies are required—and whose are not. The *fifth level* relates to the fact that the shaping (framing and formulation) of problems is a continual iterative process throughout the entire project and requires an evolving re-definition of the problem in a specific context. A "governance of problems" during the project ensures that the problem and the related purposes are not lost from sight.

The challenges on these five levels are interwoven with the definition and justification of quality criteria, most of which have to be defined at the very beginning of problem-oriented interdisciplinary projects: As Armin Grunwald (2004, 1) states, the "quality properties of integrative research depend considerably on decisions of relevance which must be made [...] before beginning the research." To date, there seems to have been too little reflection on the *problems with problems* in view of their significance for the design, course, and quality of a project.[24]

An example from science policy

The Roco–Bainbridge report of the US National Science Foundation, with its object-oriented understanding of interdisciplinarity discussed in the previous chapter, can be contrasted with an equally prominent report drawn up by a group of experts for the European Commission in response to the US initiative. The Nordmann report, entitled "Converging Technologies: Shaping the Future of European Societies," bills itself as a "specifically European approach to converging technologies" (Nordmann et al. 2004). Its core concept is "Converging Technologies for the European Knowledge Society" (CTEKS). The European group does not focus on human enhancement or the self-improvement of humankind but engages more comprehensively with societal innovation processes and with the "knowledge society." Its goal is to "expand the circles of convergences": Not the technologies themselves but the goals of technology development and research agendas are intended to converge.

Like the Roco–Bainbridge report, the Nordmann report pursues the aim of promoting "strong interdisciplinarity for research. [...] Research is needed about the processes of innovation and diffusion, the economies of artificial environments, conditions for multidisciplinary, interdisciplinary, and transdisciplinary work" (Nordmann et al. 2004, 45/41).[25] The approach taken by the Nordmann report is reflexive inasmuch as it does not simply consider interdisciplinarity as a given or as an established organizational principle having a recipe with clear milestones. Rather, interdisciplinarity is seen as posing a challenge and it creates independent research questions— research *about* interdisciplinarity and *for* interdisciplinarity, which encompasses a setting of goals.

Most relevant to our context of engagement in a *Philosophy of Interdisciplinarity* is that the two reports differ in their basic understandings of interdisciplinarity. The CTEKS concept of the Nordmann report advocates problem-oriented interdisciplinarity, although this specific term is missing. As previously mentioned, CTEKS convergence comprises more than the four technology types described in the Roco–Bainbridge report and the convergence of NBIC technologies. The Nordmann report includes an idea of convergence and also a concept of interdisciplinarity—which is called "widening the circles of convergence," namely the convergence of "nano-bio-cogno-socio-anthro-philo-geo-eco-urbo-orbo-macro-micro." One might say that the Nordmann report does not accept the narrower US understanding of "convergence."

But not only is the convergence circle wider and broader. More importantly, that which is to converge is different: "Converging technologies [of the European Knowledge Societies] converge towards a common goal" or a shared vision (Nordmann et al. 2004, 4). The Nordmann report advocates *convergences of goals*, whereas the Roco–Bainbridge report prefers a *convergence of objects* or *technical systems*: In the Nordmann report, the goals, purposes, and problems to be solved are not taken for granted; rather, they must be found and formulated. Thus, the CTEKS approach of the Nordmann report

> always [entails] an element of agenda setting. Because of this, converging technologies are particularly open to the deliberate inclusion of public and policy concerns. Deliberate agenda-setting for converging technologies can therefore be used to advance strategic objectives such as the Lisbon Agenda.
>
> (Nordmann et al. 2004, 4)

The report uses the concept of "converging technologies" as part of a discourse on our common future with the aim of determining research and development goals. Discursive, deliberative processes—associated with keywords such as participatory governance and technoscientific citizenship— are favoured, which makes apparent their proximity to discourse ethics. The Nordmann report calls for the participation not only of experts but also of citizens and the parties concerned and the incorporation of their respective knowledge and experience throughout the agenda-setting process:

> CTEKS agenda-setting is not top-down but integrated into the creative technology development process. Beginning with scientific interest and technological expertise it works from the inside out in close collaboration with the social and human sciences and multiple stakeholders through the proposed WiCC initiative ('Widening the Circles of Convergence'). For the same reason, ethical and social considerations are not external and purely reactive but through the proposed EuroSpecs process bring awareness to CT research and development.
>
> (Nordmann et al. 2004, 4)

Normativity is explicitly emphasized in the report: a "normative setting," or the setting and justification of normative elements, is at the core of the "interdisciplinary excellence" of the EuroSpecs process (Nordmann et al. 2004, 42). To achieve "interdisciplinary excellence,"

> CTEKS research programs require and produce new standards for interdisciplinary research. Interdisciplinarity usually means that researchers from various disciplines pool intellectual and technical resources as they address a problem together. This form of interdisciplinarity is insufficient when the CTEKS agenda-setting process requires critical and comparative assessments of the viability of proposals. Mutual criticism across disciplinary boundaries is required [...]. Funding incentives for collaborative research is not enough to produce this kind of interdisciplinarity.
>
> (Nordmann et al. 2004, 46)

Thus, the Nordmann report shifts the perspective away from object-oriented interdisciplinarity towards problem-oriented interdisciplinarity, which, by means of a detailed specification of each component, aims to achieve a framing of the problem, a convergence of goals, and critical reflection on and the potential revision of purposes.[26] Problems are at the centre; they constitute the starting point for research programs and individual projects alike: Such a perspective "envisions that various European Converging Technologies research programs will be formulated, each addressing a different problem and each bringing together different technologies and technology-enabling sciences" (Nordmann et al. 2004, 4).

Summary and prospect

In public and inner-academic discourse, problem-oriented interdisciplinarity is increasingly seen as a promising and even indispensable approach for addressing wicked societal problems. This form of interdisciplinarity has become known as transdisciplinarity, although a more detailed analysis could reveal that "transdisciplinarity" is used in a wider sense and is more unspecific (see Chapter 2).[27] The *Philosophy of Interdisciplinarity* strives to give substance to the scholarly jargon of "problem orientation" since the notion of "problem" is, on a conceptual level, a desideratum. From what has been outlined above, problem-oriented interdisciplinarity can be demarcated from the other three types of interdisciplinarity in the following way:

First, problem orientation versus technology-induced or object-oriented approaches: Problem-oriented interdisciplinarity does not relate mainly to technology and to technical products, nor does it address solely objects or fields of inquiry that are purely of academic interest; it is much more comprehensive. As Jürgen Mittelstraß (1992, 26) stresses,

> While the *technology-induced approach* ties in with kinds of technology in an object-oriented way [...], the *problem-oriented approach* instead links to existing and foreseeable problems and technology deficits. However, unlike the technology-induced perspective, the problem-oriented approach is not reactive; and, it is not only concerned with technology-induced issues.[28]

Initiating research by addressing societal problems is different from a technology-induced approach, as Niklas Luhmann (1998, 794) has also remarked: With the latter, "[technical] solutions continually seek [ex post] the problems that they have already solved, in order to justify their existence."[29] One therefore can draw a clear line between problem-oriented interdisciplinarity on the one hand and object-oriented interdisciplinarity, which is most predominant in the engineering sciences, on the other hand.

Second, we can distinguish between problem orientation and theory or concept orientation. As Bechmann and Frederichs (1996, 17) argue,

> problem-oriented research differs from basic research. [... Problem-oriented research] focuses on problems that occur in the societal realm; while basic research, the purpose of which is knowledge in and of itself, does not answer to any other stimulant than that of research itself.[30]

They continue, "problem-oriented research cannot wait until the fundamentals of this domain are clarified in order to then provide advice based on well-established fundamental theories" (ibid.). The traditional, strong dichotomy between theory and practice emerges here in a modified manner since problem orientation concerns societal practice. Thus, there is a difference between problem orientation on the one hand and the focus on fundamental epistemic questions on the other hand and consequently between problem-oriented and theory-oriented interdisciplinarity.

A *third* differentiation can be made between problem orientation and method orientation. This crucial difference was introduced to the debate on interdisciplinarity under the heading "problem orientation without method constraints" (Jaeger and Scheringer 1998). Although methods play a role in problem-oriented interdisciplinarity in that they serve to legitimize and justify the results—for example, in the concept of rational technology assessment or in sustainability science—they are not of central significance: *Anything goes, any method is fine*—if it serves the purpose of addressing a problem (ibid.). However, the question of whether or not a methodological foundation of interdisciplinarity is needed to improve its practice is highly disputed throughout the discourse on interdisciplinarity.[31]

Finally, I would like to add three further distinctions to illustrate the problem-oriented approach. *Fourth*, disciplinary or intra-academic scientists typically pursue long-running research *programs*, whereas researchers in problem-oriented interdisciplinarity often conduct short-term *projects*.[32] *Fifth*, the term *problem orientation* is, in a sense, a more modest and more encompassing one than the instrumentalist notion of *problem solving*. Problem orientation does not succumb to the instrumentalist deceptive illusion that problems are ultimately and finally solvable. Even Karl Popper (2000, 42) maintains that "every solution to a problem creates new unsolved problems."[33] *Sixth*, problem orientation does not necessarily imply the involvement of societal actors, lay people, or other stakeholders.

In sum, "problem" is a key term in both the political and epistemological discourse and the practice of interdisciplinarity. The objective of this chapter was to elaborate on the vague yet central notion of "problem" in order to give substance to a specific type of interdisciplinarity, namely problem-oriented interdisciplinarity. In view of their relevance to contemporary science and research, the "problems" have not been given the attention and reflection they deserve: we have problems with the "problems," particularly when it comes to reflecting on and potentially revising the point of departure of projects and research programs.

Notes

1 See also the discussion in Bogner et al. (2010) and Wickson et al. (2006).
2 We have to keep in mind that, in a strict sense, such a reference to societal problems is essential to "problem-oriented interdisciplinarity," whereas "transdisciplinarity" does not pertain exclusively to societal problems (see Chapter 2). Problem-oriented interdisciplinarity is always transdisciplinary, but the reverse does not hold.
3 An instrumentalist perspective is also present in object-, theory-, and method-oriented interdisciplinarity.
4 My translation (J.C.S.). Hence, there appears to be no *differentia specifica* between the problems dealt with within the academy, including problem-oriented interdisciplinarity, on the one hand, and an arbitrary life-world activity, on the other hand.
5 See Chapter 2 of this book.
6 My translation (J.C.S.).
7 My translation (J.C.S.).
8 My translation (J.C.S.).
9 Since often non-academic and academic stakeholders are united in a joint problem orientation, this kind of interdisciplinarity is sometimes labelled transdisciplinarity, trans-academics, or trans-science, although transdisciplinarity encompasses more than problem-oriented interdisciplinarity and is thus not very specific. See Chapter 2 in this book.

10 Weinberg (1972) does not mention the term "interdisciplinarity."
11 See the thought-provoking and intense discussion on "wicked problems" as a fundament for an "adaptive ecosystems management" (Norton 2005, 131f/159f).
12 The Office of Technology Assessment (OTA) Act of 1972 also includes the notion of "problem" (OTA 1972).
13 From a different perspective, the approach taken by Chubin et al. (1986) to "interdisciplinarity" as the "theory and practice of problem-focused research" was influential on the scholarly discourse.
14 My translation (J.C.S.).
15 My translation (J.C.S.).
16 From an analytic perspective, the boundary or border is a necessary condition for being able to speak of problem-oriented interdisciplinarity and of transdisciplinarity. Thus, to use the terminology introduced by Gieryn (1983), the notion of problem-oriented interdisciplinarity always needs some kind of "boundary work" to set up the respective demarcations.
17 "Internal" and "external" refer here to the science/research system, or in other words, to the academy.
18 Jaeger and Scheringer (1998, 11f/18) distinguish between five different types of problems.
19 There are a few exceptions, e.g., the projects carried out at the Institute for Social-Ecological Research (Becker and Jahn 2006).
20 My translation (J.C.S.). See also Pohl and Hirsch Hadorn (2007).
21 The protagonists of "social ecology," who originally stood in the critical-materialist tradition, uphold this position. Consider, for example: "Rather, there is a thing such as objective, societal problems" (Becker and Jahn 2006, 311; cp. Jahn 2013; cp. Hummel et al. 2017).
22 This holds even if we need to consider different criteria to specify what a fact is. We have not entered the post-factual society.
23 As we have also seen above with respect to the three types of knowledge (system, target, and transformation knowledge).
24 Exceptions include the approach of Bergmann et al. (2012).
25 In addition: "Interdisciplinarity should be strengthened, beyond planned or institutional collaboration, in program calls and research policies from the Commission and from the European nations" (Nordmann et al. 2004, 4). Furthermore, "CT modules should be introduced at secondary and higher education levels to synergize disciplinary perspectives and to foster interaction between liberal arts and the sciences" (ibid., 5). "Commission and Member States need to recognize and support the contributions of the social sciences and humanities in relation to CTs, with commitments especially to evolutionary anthropology, the economics of technological research and development, foresight methodologies and philosophy" (ibid., 5). Moreover, "a permanent societal observatory should be established for real-time monitoring and assessment of international CT research, including CTEKS. [...] that the Commission implement a 'EuroSpecs' research process for the development of European design specifications for converging technologies, dealing with normative issues in preparation of an international 'code of good conduct'. [...] The integration of social research into CT development should be promoted through *Begleitforschung* ('accompanying research' science and technology R&D)" (ibid., 5).
26 By widening the circles, the CTEKS approach is an attempt to overcome what Segerstrale (2000) has criticized as "the missing discourse about science and society."
27 In addition, "transdisciplinarity" is not restricted to the involvement of lay people and other stakeholders in the process of knowledge production.
28 My translation (J.C.S.).
29 My translation (J.C.S.).

30 My translation (J.C.S.).
31 Efforts are being made on various levels to carry out a methodologization of problem-oriented research by developing integrative methods.
32 Since, according to Dewey (1929, 223), "problems" are essentially linked to "uncertainty," and "uncertainty is primarily a practical matter."
33 According to Becker and Jahn (2006, 312), it is "an illusionary yet prevalent view [in the discourse on inter- and transdisciplinarity] that implementing solutions intended as answers to problems makes those problems disappear."

Interlude
On shortcomings of the instrumentalist view

Perceiving and acknowledging the existence of societal problems are indeed the first steps towards a critical-reflexive approach. However, critical reflexivity is not found in all variants of problem-oriented interdisciplinarity. The reference to problems does not necessarily involve a reflexive practice, namely an explicit reflection on problems and on how problems are produced. And it does not connote consideration of the values, underlying normative convictions, and the amalgam of metaphysical and factual aspects interlaced with a particular problem.

Shortcomings of this kind are typical of many variants of problem-oriented interdisciplinarity.[1] Many such variants advocate an instrumentalist viewpoint, signifying that, in addition to adopting a means-centred approach, they are strictly solution-oriented: The existence of, and the possibility of finding, an ultimate and benign solution is attributed to and implied by the notion of problem. Such variants presuppose a *solutionism*;[2] namely, they advocate the belief that solutions (in principle do) exist and furthermore provide a final elimination of the problem. By contrast, since there is no ultimate solution in many and the most urgent cases (e.g., global change), the critical-reflexive approach is not centred primarily on solutions. It deals with the wicked problems[3] on a deeper level. It addresses the cultural background behind the emergence of a certain problem—that is to say, the values, ontological convictions, and metaphysical presuppositions underlying the problem and its societal context. The explication of the causes of a specific problem is particularly relevant to enabling and fostering sustainable development. The point of departure of the critical-reflexive approach is the realization that, in our science-based societies, societal problems are often co-produced together with the progress of science and technology. The side effects—from asbestos and chlorofluorocarbons to nuclear waste, carbon dioxide, and the loss of biodiversity—show the inherent ambivalence of scientific/technological knowledge. The critical-reflexive kind of interdisciplinarity addresses this ambivalence. In order to contribute to thwarting new problems at their very root, critical-reflexive interdisciplinarity scrutinizes the underlying dynamics of scientific/technological advancement. In other words, emerging problems and, more fundamentally, the prevention of problems in the early phases of scientific progress are the focus.

DOI: 10.4324/9781315387109-6

When interdisciplinarity is framed from such a perspective, it does "not [stand solely] for problem-solving, but [also] for a continuous process of profound self-renewal." Jantsch (1972, 102) argues that interdisciplinarity (and transdisciplinarity) should not be regarded primarily as a better means or more efficient instrument to come up with an ultimate solution to a given problem but rather as a medium of self-reflection and self-enlightenment in order to change a situation from the bottom up. Interdisciplinarity signifies a thorn in the flesh of the academy. Specifically, it challenges the interrelation between the production of knowledge and the production of problems. Therefore, the ends, goals, or purposes of sciences or scientific projects, even of interdisciplinary projects, need to be reflected and, if necessary, be changed. Jürgen Mittelstraß (1987, 155) stresses that inter- or transdisciplinarity

> should not solely be considered as a repair initiative providing a solution that is needed when problems transcend the disciplinary scope. In complementation, inter- or transdisciplinarity—understood in the right way—serves to re-gain and to recuperate the general perception capacities and the [normative] orientation of the academy.[4]

In this sense, interdisciplinarity must be seen as an art of deeper questioning aiming to change the direction and inherent structure of scientific progress.

The critical-reflexive approach in problem-oriented interdisciplinarity is for sure an ambitious endeavour which goes beyond mere problem solving—although it is highly sensitive to problems. I will outline various aspects of this approach in the next three chapters, but before doing so I will use this short "interlude" chapter to list some further shortcomings of the instrumentalist or solutionist stance dominating many variants of problem-oriented interdisciplinarity.[5] At the heart of the instrumentalist approach is the guiding ideal that appropriate means and adequate tools need to be developed. In a nutshell, problems are taken as being given; methods and means are what matter most; an ultimate solution to the given problems is feasible; values and goals cannot—and should not—be addressed by the sciences since that goes beyond the scope of scientific rationality.[6] This position is interlaced with and fuelled by traditional dichotomies that have become engraved in our conceptualization of what is typically regarded as scientific knowledge: knowledge vs. values, facts vs. norms, and is vs. ought.[7] According to this traditional view, science is expected to tell us what the case is. Science seems to produce objective, justified, and true knowledge insofar as science is based on strict methodological guidelines; draws on empirical/experimental results; refers to a body of true propositions, concepts, or theories; and holds true regardless of human intentions and desires.

Although that type of standpoint has been vigorously disputed, its central ideal that scientific knowledge is, or must be, free of trans-epistemic values[8] is still prominent—surprisingly also in the discourse on interdisciplinarity. The thesis of value-free science was explicated by Max Weber in the early 20th century, but it originates from the birth of the modern age. In the

17th century, René Descartes provided the epoch-breaking foundation of modern science when he introduced the subject/object dichotomy, or the res cogitans/res extensa divide. His dichotomy substantiated the modern idea of the self-consciousness of the autonomous subject, on the one hand, and objective, value-free, scientific knowledge about quantifiable/describable nature, on the other hand.

Related dualisms were renewed or developed by Kant, also by Hume and Moore in the 19th century, by New Kantians like Windelband and Rickert, or by Hermeneutists such as Dilthey, Droysen, or Simmel. They postulated a big difference between academic disciplines: natural sciences vs. humanities, cultural sciences, or historical sciences. The New Kantians and Hermeneutists took the natural sciences in a seemingly value-free direction. Their viewpoint served as the basis for demarcating and defining the specific character of the humanities, including some of the emerging cultural and human sciences, in terms of understanding vs. explanation, idiographic vs. nomothetic approach, history vs. nature, or culture vs. nature. But the question then is how to characterize the social sciences—which emerged in the late 19th and early 20th century—such as economics, sociology, or behavioural psychology? These fields are strongly quantitatively, empirically, and mathematically oriented and thus not all that different from the natural sciences.

Well-known precursors of such a positivist view of the natural *and* the social sciences include Auguste Comte and John Stuart Mill. Also, Emile Durkheim and Max Weber advocated, though from different angles, the ideal of value-free empirical/experimental knowledge as a guideline for social scientists. Social scientists, they believed, should copy and mimic what seems to have guaranteed the success story of the natural sciences and of physics in particular. Science per se—which includes the social sciences—is expected to provide neutral expertise that can be used in various (e.g., good or bad) ways: Scientific knowledge production, on the one hand, and the application of scientific knowledge in societal contexts, on the other hand, appear to be strictly different enterprises.

In the following, I will look briefly at some lines of criticism levelled at instrumentalist approaches.

First, the (positivist) fact/value dichotomy (or knowledge/normativity split) underlying the instrumentalist account of interdisciplinarity has given rise to waves of criticism: from schools embracing the materialist, pragmatist, and constructivist tradition, to the early debates during the founding period of the social sciences, to Robert K. Merton's later seminal work in the 1940s[9] and the positivist dispute in the 1960s, to more recent movements such as new experimentalism, social or cognitive constructivism, social epistemology, science and technology studies, or feminism. The critics argue that a (positivist or neopositivist) value-free understanding of science and of scientific knowledge is just a myth. It is too limited, too decontextualized, and too simplified to correspond to the practice of the scientific enterprise. Some critics go so far as to object to the so-called naturalistic fallacy, which maintains that there is no transition from *Is* to *Ought*. Acknowledging a fallacy in this regard means presupposing a dichotomy between the two spheres (cp. Jonas 1984, 44). Critics argue that if such a limited under-standing of scientific knowledge—and of the *Is* as a mere "fact"—were to be exposed to a deeper analysis, the fact/value dichotomy, and what was branded a fallacy, would turn out to be nothing but a circular thinking cycle or a meaningless tautology. Pragmatists in the Anglo-American tradition have objected to ascribing the fact/value dichotomy to an ontological or any other fundamental level and, in consequence, have turned it into a mere heuristic that is interlaced with the aim being pursued (Dewey 1929).

In addition to Jürgen Habermas's (1971) arguments, which were drawn up at the advent of the positivist methodology dispute in the social sciences, Hillary Putnam (2002) in *Collapse of the Fact/Value Dichotomy* articu-lates a more recent rejection in connection with the myth picturing sci-ence as value-free. Putnam objects to any ontological dichotomy but not to a pragmatic context-specific distinction.[10] As he notes, "the fact/value dichotomy is, at bottom, not a distinction but a thesis" that is inflated with "metaphysical" contents (ibid., 19). Essentially, it can be "defended [only] on metaphysical grounds" (ibid., 40). Therefore, we can conclude, any kind of instrumentalism—even one that purports to be anti-metaphys-ical—is interlaced with metaphysical presuppositions. Besides Putnam, the field of Science and Technology Studies (STS) has addressed various fact/value hybrids, boundary objects, and trading zones (Hackett et al. 2008). This field has strongly questioned these dichotomies, viewing them as mere academic constructions or normative ideals that cannot be justified by ref-erence to the practice of the scientific enterprise.

From a descriptive perspective, the fact/value dichotomy is therefore only a delusion. Moreover, from a more normative angle, it is debatable whether the dichotomy provides orientation in the world we live in. Some scholars claim that the fact/value dichotomy should be preserved, at least heuristi-cally, in order to guarantee action-theoretical approaches and to correlate three aspects of human action: (1) goals/intentions, (2) means/instruments, and (3) consequences/results. But is their claim based on sound arguments? It may well be the case that the opposite is true: The fact/value dichotomy might foster ignorance and blindness about what is at stake with regard to the unsustainability of late-modern knowledge societies at large.

Let us now consider, *second*, another line of criticism: the critical-materialist or transcendental-pragmatist one. Critical materialists perceive the fact/value dichotomy as the result of a societal process of erasing the value, purpose, and goal perspective. The market-driven dynamics of the technoscience-based knowledge industry induces a loss of ends—and the replacement of ends with means, methods, and algorithms. Through this process, ends and goals are eliminated, concealed, or excluded from that which counts as knowledge: They are stigmatized and devaluated as being contingent and subjective, whereas scientific knowledge or facts are deemed to be objective and based on clear evidence and truth criteria. This prevalent dichotomy is thus a result of the market-driven dynamics of the modern age, witnessed even in the university and research system, towards a means-, method-, and algorithmic-oriented rationalization. According to Max Horkheimer, Theodor W. Adorno, and Herbert Marcuse, the "subjectivization of ends" is part of an ambivalent historical process of a formal rationalization and secularization of societies in general.

The subjectivization of ends has become institutionalized in the alienating production conditions of the capitalist economy and its leading institutions, which include contemporary, neoliberally driven universities. Critical theorists argue that the dichotomist means/ends rationality has undermined the core ideal of the Enlightenment: reason and its inherent critical potential to scrutinize and question what is given. Over the course of history, reason has been transformed and reduced to what Horkheimer branded "instrumental reason," thereby revealing the dialectic inherent in the tradition of the Enlightenment. Means/ends rationality is today becoming increasingly prevalent and governing individual, social, and institutional actions. It cements the unequal and unjust distribution of power in society at large.

In light of the tendency to lose or abandon ends and to disregard purposes, critical materialist philosophers strongly resist throwing out the baby with the bath water. They put their focus on maintaining the normative aspect of action theories while placing strong emphasis on ends, goals, or purposes and seek to re-establish and institutionalize a participatory, rational discourse on ends (Habermas 1970, 1984). They see reflection on and the potential revision of ends as being central to a rational societal as well as scientific discourse and demand that ends be deliberately defined by informed public consent via a power-free discourse among equally informed, communicative actors—unswayed by interest groups such as neoliberals or other stakeholders. Many of these ideas have been condensed in Karl-Otto Apel's and Jürgen Habermas's transcendental pragmatist *Discourse Ethics* (Apel 1988; Habermas 1993).

Third, whereas critical materialist and transcendental pragmatist philosophers uphold the possibility and the effectiveness of a rational discourse on ends, purposes, and problems, environmentalists and environmental philosophers typically take a different approach. Since the emergence of major environmental problems in the 1960s, environmentalists have perceived these pressing issues to be symptoms of a more fundamental (cultural or knowledge) crisis: of the predominant way in which society perceives, conceives, and frames nature and the societal relations to nature, including the guiding ontologies and metaphysical concepts of nature. The environmental problems cast a shadow on what the modern epoch has stylized as a core achievement: scientific knowledge about nature. The sciences themselves—the production of scientific knowledge and the instrumental shortcomings resulting from the means/ends split—are at stake here. According to environmentalists, global change problems cannot be seen as side effects of technoscientific progress that are eliminable in principle or as having been induced simply by the inappropriate application of technology by certain stakeholders. Rather, the problems are intrinsically intermingled with the modern way of framing nature and the societal relations to nature and of conceptualizing knowledge about nature. Environmentalists have identified and branded a technomorphic way of thinking of nature—including what is prevalent in the instrumentalists' account—as the central underlying source of the problems. Since scientific/technological progress can no longer be equated with societal/human progress, the Baconian age of value-free knowledge and instrumental reason comes to a close, Gernot Böhme (1993) argues. In line with the criticism raised by environmentalists, Georg Picht (1969, 80) disputes what is typically regarded as knowledge: "The present-day kind of knowledge that is interlaced with the destruction of its objects—in other words: that destroys nature in technological apparatus and in daily technical actions—cannot be considered as true." Besides criticizing Francis Bacon, Picht accuses René Descartes for his alienated view of nature and his strong ontological dualism. Like Picht, Hans Jonas questions the value-neutrality thesis, arguing that

> if the picture that the natural sciences portray of nature were the ultimate word on what is the essence of the whole world, the latter would be a value-neutral mechanical gear. [...] Men would have no duty to care about nature.[11]
>
> (Jonas 1993, 44)

Arne Næss (1973) and the deep ecology movement even go so far as to posit the need for a cultural shift in the conceptualization of nature towards a perception that is linked to a kind of spiritual thinking. In sum, many environmental philosophers diagnose the origin of the environmental crisis as lying at the very beginning of the modern epoch, during the time of Bacon and Descartes with their specific materialist concept of nature as a mechanical system which became the culturally predominant interpretation: We are faced not solely with an environmental crisis but also with a cultural, societal, or scientific one.

A further point of criticism, similar to the one outlined above, is, *fourth*, articulated by Martin Heidegger as well as by phenomenologists and cultural philosophers. In his work *The Question Concerning Technology*, Heidegger argues that the instrumental and anthropological, means-oriented framing of objects of the entire world (and, therefore, of nature) is not solely central to modern technology but that this kind of thinking and framing can already be found in the sciences:

> Modern science's way of representing pursues and entraps nature as a calculable coherence of forces [and as mathematical laws]. Modern physics is not experimental physics because it applies apparatus to the questioning of nature. Rather the reverse is true. Because physics, indeed already as pure theory, sets nature up to exhibit itself as a coherence of forces calculable in advance, it therefore orders its experiments

precisely for the purpose of asking whether and how nature reports itself when set up in this way [...]. The modern physical theory of nature prepares the way not simply for technology but for the essence of modern technology. [...] Because the essence of modern technology lies in *Enframing [in German: Ge-stell]*, modern technology must employ exact physical science. Through its so doing, the deceptive illusion arises that modern technology is applied physical science. This illusion can maintain itself only so long as neither the essential origin of modern science nor indeed the essence of modern technology is adequately found out through questioning.

(Heidegger 1977, 21f)

For Heidegger and the phenomenological tradition, the instrumentalist approach and its technomorphic thinking have been initiated and fostered by the exact sciences: Technomorphic thinking has become crystallized in modern physics and also in modern action theories.[12] Therefore, the *Crisis* of modern science—to paraphrase Edmund Husserl—consists in the way in which the sciences approach and frame nature and humans' relations to nature: The crisis is "rooted in the abstraction by which [... science] views the life-world as just an ensemble of bodies" (Husserl 1950, Series 6, 230).[13] From Heidegger's diagnosis of "enframing" and Husserl's of "abstraction," we can conclude that, in order to change our behaviour and action today, we need to rethink our thinking and to reframe our framing of nature and humans' relations to nature. To enable such a cultural transformation, we need to take a closer look at exact science—since its conceptualization of nature also determines our understanding of nature in our life-world and our day-to-day practices concerning nature. These concepts and preconcepts strongly matter.

The above-listed four lines of critique have been articulated from a perspective "external" to the sciences. In addition, we need to consider, *fifth*, that critical voices are also raised from an "intra-scientific" perspective. These voices complement the attempt of environmentalists and phenomenologists to frame and understand nature in a different way. They argue in favour of opening avenues towards a pluralistic and more differentiated understanding of nature and of humans' relations to nature *and* of scientific knowledge. The Belgian Nobel laureate Ilya Prigogine and the philosopher Isabelle Stengers criticize the traditional exact sciences for having largely failed to acknowledge and to address nature's temporal, dynamic, evolutionary side. In their *Order Out of Chaos: Man's New Dialogue with Nature*, Prigogine and Stengers (1984) advocate a fundamental transformation in the conceptualization of nature, namely *From Being to Becoming* (cp. Prigogine 1980). Their basic objective is to facilitate an anti-reductionist and anti-mechanistic *participatory view of nature* that is based on dynamics, self-organization, temporality, instability, and complexity. Today, such a view is shared by those scholars of the environmentalist tradition who see the environmental crisis as a deeper one, namely as a cultural, institutional, and knowledge crisis.

Prigogine and Stengers argue that, since the late 1960s, the way in which nature is seen has been undergoing a structural paradigm shift which, *first*, reveals the limitations of the modern concept of nature and, *second*, enables the integration of various knowledge fragments into a novel picture of nature. At issue is also, *third*, the modern understanding of knowledge, which can be traced back to Plato's thinking. Although the criteria defining scientific knowledge have been subjects of major dispute, scholars have reached a consensus that the central criteria encompass predictability, reproducibility, testability, and explainability (cp. Schmidt 2011a, 2015a). According to Prigogine and Stengers, these criteria are now undergoing modification since they are based on mistaken assumptions about nature and the natural objects under consideration—specifically, the assumption of stability: Framing nature through the lens of stability is erroneous. To give substance to their claim, Prigogine and Stengers point to recent advancements in the theory of dissipative structures and nonlinear thermodynamics far from equilibrium and also in general to nonlinear dynamics, complex systems theories, chaos and catastrophe theory, synergetics, autopoiesis theory, fractal geometry, and the like. These interdisciplinary concepts are based on instabilities—as will be shown in Chapter 7: In nature as well as in the social sphere, instabilities turn out to prevail as the main source of self-organization and evolutionary processes. Instabilities exhibit sensitive dependence on initial or boundary conditions. This observation supports Prigogine's argument that the stability assumption made in traditional sciences is nothing but a metaphysical presupposition or a "dogma." Despite there being some rationale behind it throughout the history of science, the stability presupposition inhibited progress with regard to understanding nature in the 20th century (cp. Schmidt 2015a).

Today, instabilities prompt criticism of the well-established criteria defining (and defending) knowledge: predictability/calculability, reproducibility/experimentation, testability/confirmability, and describability/explainability. Weaker criteria are gradually replacing the rigorous requirements that previously qualified knowledge as *scientific* knowledge. A novel view of science that can be called "late modern" is emerging (Schmidt 2008a, 2011a) and may serve as an example of a critical-reflexive version of interdisciplinarity. This kind of science enables self-awareness, self-critique, and self-reflexivity. Interdisciplinarity emerges within the scientific disciplines, when and if disciplinary boundaries are transcended and societal problems are addressed in a critical-reflexive process. Framed from this angle, disciplinarity and interdisciplinarity are not mutually exclusive or contradictory but instead go hand in hand.[14]

In sum, the critical-reflexive approach of problem-oriented interdisciplinarity does not view inter- and transdisciplinarity primarily in terms of the problem–solution schema consisting of (a) goals/intentions, (b) means/instruments, and (c) consequences/solutions. It does not fall into the trap of *instrumentalism* or *solutionism* with their respective shortcomings, namely their belief that the challenges of global change and the threats to the environment in the Anthropocene ultimately can be managed and mastered. More fundamentally, the critical-reflexive approach of problem-oriented interdisciplinarity aims to address problems on a deeper level of our culture. In the following sections, some pathways towards a critical-reflexive perspective and case studies will be presented.

Notes

1 My approach shares much with that of Becker and Jahn (2006), Jahn (2013), Frodeman (2010), Frodeman (2014), and Hummel et al. (2017).
2 This is a notion taken from Morozov (2013) but used here with slightly different connotations.
3 Consider, for instance, the definition of "wicked problems" by Rittel and Webber (1973) (see previous chapter).
4 My translation from German (J.C.S.).
5 The instrumentalist bias can also be shown for the object-, theory-, and method-oriented types of interdisciplinarity.
6 For a critique, see Latucca (2001), Holbrook (2013), and Frodeman (2014).
7 This includes what was branded the "naturalistic fallacy."
8 See Chapter 2 in this book: The thesis of value-freeness states that only epistemic values (e.g., empirical correspondence, consistency, coherence, explainability, objectivity, and fruitfulness) have to play a role in the sciences. Other (non-epistemic) values such as economic, social, religious, or personal ones must be excluded.
9 See, as a synopsis, Merton (1973).
10 Putnam (2002, 133) stresses—by referring to the ideas of Habermas—that "ethical values can be rationally discussed."
11 My translation (J.C.S.). See also Chapter 6.
12 Instrumental or means/ends rationality does not only arise through the sciences; it is also an ambivalent precondition that is highly interwoven with the sciences.
13 My translation (J.C.S.).
14 This is in line with von Hentig's (1972) approach, in which he identifies an interdisciplinarity-focused "good or sound disciplinarity."

6 Ethics and the environment

Engaging with grand environmental challenges of the cultural crisis

Environmentalist concept

Interdisciplinarity originally emerged in environmentalism, which came up in response to the destruction of the natural environment.[1] Besides pioneers like Erich Jantsch (1972), the philosopher and ethicist Hans Jonas addressed the challenge.[2] His environmental philosophy advocates a critical-reflexive type of problem-oriented interdisciplinarity—although he does not use these notions explicitly. Jonas questions the way in which present-day science conceptualizes and treats nature. He argues that a fundamental transformation of the recent dominant concept of nature is necessary. According to Jonas, such a change will not be accomplished by an instrumentalist approach. The latter is too weak given what is at stake.[3]

Jonas's future-oriented, anti-visionary environmental philosophy and ethics, including his philosophy of nature, had a tremendous impact on public and philosophical debates throughout the 1980s and the early 1990s. He has been credited with taking the recent situation of our natural environment, our life-world, and our sociotechnological culture seriously. His thinking focuses on pressing environmental problems, in connection with the ambivalence of scientific and technological advancement. His approach therefore matches perfectly with what we have discussed so far under the heading of problem-oriented interdisciplinarity—and with the *Philosophy of Interdisciplinarity*. Jonas does not advocate a purely academic, disciplinary philosophy appealing to and reaching just a limited audience of experts: Good philosophy, he maintains, has to face—in one way or another—the present-day challenges of society at large. Philosophy needs to be good for something; it has a general duty to contribute to the survival of the human species and the future of humankind: "Care for the future of mankind is the overruling duty of collective human action in the age of technical civilization that has become 'almighty'" (Jonas 1984, 136). His environmentally focused, catastrophe-prevention "emergency ethics" encompasses a precautionary principle and, accordingly, a "'heuristics of fear', replacing the former projections of hope" (ibid., x), whereas the latter has been—since Francis Bacon's pioneering but ambivalent works such as *New Atlantis* and his *Novum Organon*—central to the modern epoch.

DOI: 10.4324/9781315387109-7

Though severely criticized by the scientific community of academic philosophers, Jonas's politically as well as philosophically influential book *The Imperative of Responsibility. In Search of an Ethics for the Technological Age*[4] has been a bestseller. It was a forerunner of the idea of sustainable development, the UN Sustainable Development Goals (SDGs), and various political programs of socioecological transformation.[5] His philosophically grounded and interdisciplinarily oriented work also set off an impressive wave of incorporating science and engineering ethics in university curricula of disciplinary engineering and science education programs. Jonas is a pioneer in driving the idea of critical-reflexive interdisciplinarity—including a synthetic-synoptic dimension drawing on his philosophy of nature and philosophy of biology—in order to shift the direction of scientific advancement onto an environmentally friendly path. The call for a fundamental reorientation is interlaced with some programmatic and highly disputed ideas. As Jonas wrote in the 1980s, we need several "new sciences which have to deal with an enormous complexity of interdependencies," an interdisciplinary or "integral environmental science" (Jonas 1987, 11f).[6]

Roots of the crisis

By addressing and rethinking the background to the environmental crisis, Jonas does indeed advance strong claims and provocative arguments. His diagnosis that we are confronted with an intrinsic ambivalence of technoscientific advancement, formulated as early as in the late 1970s, was based on the perception that we "live in an apocalyptic situation"—facing a crisis with various "threats" (Jonas 1984, 140/ix). The threats are not caused solely by the unintended side effects of technology, including the misuse of technical systems, but rather by the achievements of science, engineering, and technology themselves. According to Jonas,

> it is the slow, long-term, cumulative—the peaceful and constructive use of worldwide technological power, a use in which all of us collaborate as captive beneficiaries through rising production, consumption, and sheer population growth—that poses threats much harder to counter. The net total of these threats is the overtaxing of nature, environmental and (perhaps) human as well.
>
> (ibid., ix)

The emerging environmental crisis, Jonas argues, is deeply rooted at the metaphysical fundament of contemporary culture and is interlaced with its predominant way of perceiving and conceptualizing nature, which includes the separation and segregation of scientific disciplines into silos. Nature is becoming widely disposable and questionable. At the same time, it is becoming radical on the subject of humans' actions and projects.[7] This overtaxes traditional ethics: Ethics alone can no longer cope with the ensuing challenges. Moreover, the crisis challenges the metaphysical fundament of Western culture and calls for a revised view of nature ("metaphysics") and a change in the underlying interrelationship between ethics, philosophy of nature, and anthropology. Thus, the crisis "push[es] the necessary rethinking beyond the doctrine of action, that is, ethics, into the doctrine of being, that is, metaphysics, in which all ethics must be grounded" (Jonas 1984, 8):[8] "[M]etaphysics must underpin ethics" (ibid., x). Major challenges require fundamental reflection on and (potentially) a revision of the metaphysical bases and ontological backgrounds. Anything else is too weak, so thought Jonas 40 years or so ago.

In the present day, however, short-term instrumentalist or strategic thinking seems to dominate public policy: quick regulations in response to pressing problems. Jürgen Mittelstraß (1987) speaks of a "repair ethics" and a repair approach taken by the associated type of instrumentalist interdisciplinarity: Ethics is obviously far too reactive and hardly proactive or prospective. Seldom is it more than cosmetics and compensation. Yet global change issues—along with advances in synthetic biology, CRISPR/ Cas, nanotechnologies, medical/neuro-enhancement technologies, or artificial intelligence—renew the call for a much more deeply rooted ethics of nature and technology: Is it not a permanent obligation of an interdisciplinary-oriented philosophy to render the underlying metaphysical assumptions open to rational reflection and hence make them revisable? Would philosophy not be well advised to express its close relationship to the life-world and to society in programmatic terms: *shaping concepts and views of nature ("metaphysics"), underpinning ethics, building society and culture?*[9] In short, what kind of ethics, philosophy of nature, and anthropology do we need in order to progress *from* intra-philosophical justification discourses *to* philosophically reflective action and decision-making in the life-world and in society?

Certainly, such thinking is provocative today, although these programmatic theses—prominently presented in Jonas's *Imperative of Responsibility*—were broadly debated in the 1980s. Christoph Hubig (1995, 13) calls Jonas's book a "philosophical bestseller." It contributed in a major way to the boom of philosophical ethics in the 1980s. Ethics sought to be more than (and different from) a "bicycle brake on an intercontinental airplane," to cite the reproach directed at ethics by Ulrich Beck (2007, 73f). Ethics went public and was perceptible and audacious; it engaged and got involved. One expected it to contribute to societal development and to deliver something, such as guiding principles and answers for our societal futures or, in other words, a non-formal, normative knowledge capable of providing orientation. Jonas's public philosophy therefore can be regarded as an "engaged philosophy," a term coined by Robert Frodeman (2010) to foster a re-orientation of disciplinary philosophy. Such a philosophy can be regarded as interdisciplinary in a (self-)critical-reflexive sense—an interdisciplinary philosophy that is part of any good reflexive and reflective practice.

Jonas's late work, which was pivotal to his third thematic period following his investigation of gnosis and his philosophy of organisms, has earned an eminent status in the history of philosophy and also in cultural philosophical terms. His *Imperative of Responsibility* can be described as being both product and promoter of societal and cultural reflection; to a certain extent, it is a mark of a "reflexive modernization" (Beck 1992). Jonas's publication represented a socio-historical milestone—or even caesura—which contributed to a reconsideration of the societal relations to nature underlying the socio-ecological crisis; Jonas comprehended the crisis as a challenge to philosophy.

Now, though, it appears that this socio-historical transition into the "reflexive modern society" (Beck 1992) is infinitely remote to us. So much has the transition become taken for granted that every trace of questionableness seems to have vanished, leaving behind it an amorphous, unquestioning acceptance. No other seems to have become, like Hans Jonas, such a symbol of an evidently concluded epoch—of an overcome alarmism, stimulated by moralizing and metaphysically religious "good human beings" in conjunction with cultural pessimism and conservative-naturalistic talk of crisis and apocalypse prevention in the face of alleged threats of doom. Former normativity seems falsified by present-day normality. Debates on *risks* are regarded as obsolete. Speaking about an "ecological catastrophe" is an exaggeration, or so it seems to those who criticize Jonas; genetic engineering and medical technology are lagging behind their utopian expectations; the *New Human* is not in sight; societal reality, individual life-world, and human beings' biological constitution all still appear stable enough.

Postmodernism, pluralization, and particularization also played their part in establishing a purported consensus about a radically post-metaphysical age (Habermas 1992). Deconstructivism and constructivism dispelled the "gravity" of a problem-oriented objectivity—arguably exaggerated from an "alarmistic" perspective—and portrayed the problems themselves as socially constructed: The hole in the ozone layer disappeared as a social phenomenon in the contingency of plural perspectives of socio-cogno constructions. The "power program" thinker Francis Bacon, who set out to "conquer" nature, has also made a comeback. As a reflection of the societal climate, Jonas's approach seems to have vanished from philosophy. A normative anti-normativism has emerged; a value-free description of allegedly non-referential phenomena of societal reality seems feasible *in* and *via* philosophy of all things—not an opportune time for Jonas's approach of putting central focus on reflecting on and potentially revising metaphysical considerations.

In light of recent issues such as those linked to global change, it might nevertheless be appropriate to reconsider whether Jonas's approach of establishing an interrelation between ethics, philosophy of nature, and anthropology, and his argument for a future-oriented responsibility on a global scale might not, after all, contribute to a critical analysis of the present state of affairs and of our societal relations to nature. At the same time, Jonas's thinking is paradigmatic for a critical-reflexive account of interdisciplinarity. In the following, the main objections raised against Jonas will be reconstructed and critically assessed.[10] It will be shown that most of the objections are not as sound as the critics claim them to be—although Jonas's argumentative justification is somewhat weak. Specifically, the impact of Jonas's future-directed and problem-oriented approach on the recent debate on sustainability will be underlined. His "Imperative of Responsibility" has been adopted by the Brundtland report and is prevalent in various concepts of sustainable development (WCED 1987). A major objective of this chapter thus is to show how Jonas regards "ethics" and "philosophy of nature" as twin sisters or, more precisely, how "ethics becomes part of the philosophy of nature."[11] Since a philosophy of this kind faces, in a fundamental sense, real-world problems, it can serve as an illustrative example of the critical-reflexive account of the *Philosophy of Interdisciplinarity*.

Objections

Let us begin with the critics. What did they hold against Jonas? Why did they believe that Jonas's approach is nothing but a dead end? In the early 1990s, Lothar Schäfer maintained that "we should not, and indeed must not, follow the propositions made by Jonas" (Schäfer 1999, 96; cp. Schäfer 1993)—a view which soon seemed to take hold in philosophy and in public. Schäfer accused Jonas of delivering merely "religious answers" to dilemmas that are not religious but societal (Schäfer 1999, 28).[12] For Schäfer, the essential task is to reshape the "Bacon Project" (as Schäfer's book is entitled), to reformulate it for the future under modified conditions, and to undertake a critique of recent political economy.

Another major point of criticism concerns significant problems surrounding the application and implementation of his approach. According to Angelika Krebs (1997, 379) those like Jonas "who press for a radical change, a paradigm shift, in our moral attitude toward nature, and brand anthropocentrism as the source of all ecological evil [...] are [...] wrong" and do not help to solve the problem.[13] Gertrude Hirsch Hadorn (2000, 385) discerns in Jonas's ethics a problematic "secularized belief in God" and a "classical ontology" with a "philosophy of the absolute" leading to eminent "problems of justification" that "raise doubts about his philosophical responses to these questions" (ibid., 114/384/377): "Jonas' Imperative of Responsibility," Hirsch-Hadorn writes, is "onto-theological moral grounding without a moral definition. It leaves the interpretation of the orientation problem of technological civilization [...] to arbitrariness" (ibid., 387).[14] Heiner Hastedt (1994, 178/170) identifies a "tendency in Jonas' naturalistic ecological ethics to formulate in a vehemently moralizing way general and addressee-less postulates" that miss the "societal dimension of the problem" and moreover themselves tend to exhibit "inconsistency" in their arguments.[15] Armin Grunwald (1996, 200) believes that Jonas's reasoning is untenable analytically and that his ethics is socially irrelevant. "Philosophical attempts to establish a 'duty of mankind to exist' are therefore not only problematic because of metaphysical presuppositions, but also superfluous."[16] For Grunwald (1996, 202f), the future path of an ethics of technology [...] is predestined: away from metaphysical justifications, an increasing liberation from Jonas' legacy of an 'apocalypse-prevention ethics', [...] towards situative decision logic for the individual case [... and] towards a pragmatism of discursive conflict management.[17]

In addition, other critics focus on the precautionary principle, which is strongly supported by Jonas in various forms ("in dubio pro malo").[18]

Some of the objections raised against Jonas have thus far been outlined: *First*, the *diagnosis objection* critiques Jonas's perception and diagnosis of a cultural crisis and the associated epochal break thesis. *Second*, the *origin analysis objection*, though it may share Jonas's diagnosis or motives, does not identify any fundamental challenge to ethics, metaphysics, or philosophy in general. *Third*, the *argumentation and justification objection* accuses Jonas of fallacies, incorrect deductions, and unfounded arguments. *Fourth and finally*, the *practice objection* levels the accusation of ineffectiveness and irrelevance.

The criticism, it seems, is overwhelming. Nevertheless, one might ask, are the objections against Jonas as well founded and sound as they evidently appear to be? In the following, I present a critical and analytical reconstruction of the objections to Jonas's argument of "responsibility for the future."[19] I argue that—despite some deficits in its context of justification—Jonas's overall argument does exhibit a high degree of plausibility.

Diagnosis—and against the first objection

Jonas (1984, 140) asserts that "we live in an apocalyptic situation, that is, under the threat of a universal catastrophe if we let things take their present course." He identifies an epochal break in the history of humankind. The Baconian ideal of infinite progress has given rise to a threat—not only because of unwanted and unintended side effects but, more fundamentally, because of the excessive magnitude of its success. The existence of human beings as a biological species and the human–societal existence of future generations are at stake. The "original venture [of freedom]" prefigured in the organic draws attention (through man's deeds) to the possibility of not-being: the "threat of its negative" (Jonas 2001, 5/4). That life is "an experiment with mounting stakes and risks which in the fateful freedom of man may end in disaster as well as success" (Jonas 2001, x/xxiii f)—a "progressive scale of freedom and peril"—is nowhere clearer than in the large-scale technoscientific "experiment with nature."

Jonas sees culturally established concepts of humankind, whether they be individual or societal, eroded by the "altered nature of human action, with the magnitude and novelty of its works" (Jonas 1984, x). Nature—conceived of as man's corporeal nature, as the growth of organic existence, as an external resource ("environment"), and as phenomenal, aesthetic nature—is becoming completely subject to disposal.

The subjugation of nature with a view toward man's happiness has brought about, by the disproportion of its success, which now extends to the nature of man himself, the greatest challenge for the human that his own needs have ever entailed. Everything about it is novel and unprecedented in character and magnitude. [...] That the promise of modern technology has turned into a threat forms the premise [of this book] for Jonas.[20]

A negative dialectic of freedom in our socio-technical practices emerges. Jonas discusses scientific progress from pre-modern *techne*, which basically revolved around instruments, tools, and methods, to modern technology. Mechanics, chemistry, electricity, information and communications technology, and bio-engineering have successively established themselves. Advanced technology potentiates nature's disposability and its "vulnerability" (Jonas 1984, 6). The Aristotelian distinction between nature (*physis*) and *techne*—programmatically eliminated by Francis Bacon at the birth of the modern age—is becoming obsolete (cp. Schmidt 2011b). *Homo faber* is triumphing over *homo sapiens* (Jonas 1984, 9f). For Jonas, the boundary between nature and city (*nomos, polis*) serves as a guiding metaphor to underline his diagnosis. This boundary has become blurred and obliterated as a "universal city" takes shape. "The difference between the artificial and the natural has vanished"; free nature is being "swallowed up in the sphere of the artificial" (Jonas 1984, 10). An ambivalent momentum in the advancement of technology has ensued, with cumulative impacts and evolutionary hazards, revealing the "vulnerability of nature" and the fragility of natural equilibriums. A negative dialectic of power over nature arises because "every attempt to break the natural thralldom, because nature is broken, enters all the more deeply into that natural enslavement," as noted by Max Horkheimer and Theodor W. Adorno in a tenor very similar to that of Jonas (Horkheimer and Adorno 1972, 13).

Jonas thus starts—like Georg Picht, Walter Schulz, and Carl Friedrich von Weizsäcker—with the anti-utopianist and anti-Blochian shock that being (of humans, namely the "Sein") is no longer a given and can no longer be taken for granted. It is true that this shock over the "enormity of the situation" was not, as the critics of Jonas's approach point out, accompanied by specific reflection on the plurality of technology, disparity of agents, diversity of cultures, and complexity of societal conditions, but Jonas's diagnosis nevertheless can be granted a certain degree of plausibility.

Analysis—and against the second objection

Apparently, we are incapable of taking appropriate action in the face of such "threats." Jonas characterizes the untrodden ground of collective technological practice that we have embarked on with high technology as an "ethical vacuum" (Jonas 1984, 23) and a "no man's land for ethics."[21] Conventional ethics needs to be supplemented. (1) Previously, all dealing with nature in the form of craftsmanship (*techne*) was considered "ethically neutral" insofar as nature could be assumed to be largely independent of man's actions. (2) Only human beings' interactions with other human beings were of ethical significance. (3) Human beings, as the prerequisite for ethical action, could be assumed to be a constant. (4) Global notions of space and time were not taken into account; conventional ethics was about "the here and now" (Jonas 1984, 5) and confined to the "proximate range of action": "The short arm of human power did not call for a long arm of predictive knowledge" (Jonas 1984, 6).

The critical assessment presented by Jonas is not restricted to any particular, well-established concept of ethics. It applies equally to deontology, utilitarianism, contractualism, virtue ethics, and ethics of compassion and today also to discourse ethics. As Jonas stresses, he does not seek to replace other ethical concepts but to complement and expand them.[22]

For Jonas, the ethical vacuum is not at all surprising—it is caused by an inadequate metaphysics. He identifies deficits in the metaphysical fundament of contemporary culture, by which he means the set of implicit understandings about man, nature, knowledge, and science; he is not alluding to the grand metaphysical systems in the tradition of German Idealism.[23] If it is correct to say that even secular late-modern societies are influenced by a metaphysical fundament and ontological background—and therefore we have indeed not entered a "post-metaphysical age" (Habermas 1992)—this prompts the question of whether we need to revise the metaphysics, rationalities, and underlying convictions governing those very perceptions of nature, man, and science. Fundamental crises make it necessary, Jonas writes, to "push the necessary rethinking beyond the doctrine of action, that is, ethics, into the doctrine of being, that is, metaphysics, in which all ethics must ultimately be grounded" (Jonas 1984, 8):[24] Whether and why humankind should exist are metaphysical questions. Thus, "metaphysics must underpin ethics" (ibid., x).

Metaphysics is not to be understood as a contingent positing or as a dogma immune to revision. On the contrary, in light of the tradition of philosophy since the ancient Greeks, metaphysics "has always been a business of reason" (Jonas 1984, 45).[25] Therefore, the question is not whether we have, or want, a metaphysics, but which one it can and should be. Jonas seeks to engender through rational thinking a metaphysics that is adequate to the recent threat to the survival of mankind—in brief, he advocates a pragmatic shaping of metaphysics. Recently, Jean-Pierre Dupuy, in line with Jonas, has also spoken of the need for "an adequate metaphysics." He has in focus the "foundations of a metaphysics adapted to the temporality of catastrophes" (Dupuy 2009, 10).

How relevant is the metaphysical? Metaphysics always surrounds us. The natural and engineering sciences exhibit metaphysical thinking in their approach to nature, in their methodology, and in their concepts of knowledge. "If the picture that the natural sciences portray of nature," Jonas says,

> were the last word on the condition of the world, the latter would be a value-neutral mechanical gear [...]. If this picture is an adequate

description of reality, there is indeed no reason why we should worry about the coming millennium.[26]

(Jonas 1993, 44)

According to Jonas, "nature" here is constituted and understood in a reductionist sense as an external, self-referential, positivistic matter of fact having no relation at all to man; it is seen as a static, stable, a-temporal entity. Max Weber, speaking in the same vein as Jonas, famously coined the words "dis-magnification" and "disenchantment of the world"; Horkheimer and Adorno (1972) identified a "reification" and "objectification" of nature.[27] The disenchantment induces a de-valuation and a "dispersal and involution of value whose upshot for us is total confusion," Baudrillard (2003, 10) states (cp. Höffe 1993). Jonas opposes the value-neutralization of nature and insists "that natural science may not tell the whole story about nature" (Jonas 1984, 8). Hence, he advocates an *approach thesis* and suggests that we should reflect on and potentially revise the metaphysical assumptions underlying the one or other approach. This thesis concurs entirely with Goethe: "The conclusions of men are very different according to the mode in which they approach a science or branch of knowledge; from which side, through which door they enter" (Goethe 2006, xxxi).

The approach argument transposes ethical considerations to the earlier stages of the process of knowledge production—to phases of framing, understanding, and gaining insight and in particular to the motives and attitudes of scientific activity and the constitution of research programs. In contrast, the emergence of problems has always been accepted as a matter of fact in the various branches of ethics; the underlying dynamics that induces the emergence of new problems has not been questioned or addressed.[28] This deficit, namely ignoring fundamental questions about knowledge and problem production, needs to be addressed and eliminated according to Jonas. Before applying ethics and before considering the well-established questions of ethics, we need to ask: How should we think?—as Martin Heidegger (1977, 40) put it. In essence, the way we think will influence our decisions and actions.

Argumentation—and against the third objection

The tasks of and the requirements to be met by ethics and practical philosophy have always been a matter of debate. Jonas was criticized for not having presented an argumentative grounding for his future-oriented ethics. However, this objection fails if it is intended as a general one. Despite an obvious lack of argumentative stringency at some points, Jonas does voice clear premises and definitive conclusions.

First, Jonas's argumentation is based on a teleological approach which constitutes a central part of his philosophy of nature and philosophical anthropology—and complements his ethics (*purposiveness thesis*). Of seminal significance is his main philosophy-of-nature work *Organism and*

Freedom (1966), in which he develops an integral-monistic philosophy of life encompassing his philosophy of organism and philosophy of mind (Jonas 2001). Jonas invokes Aristotle, critiques Descartes, opposes Kant, and shows an affinity with Schelling and also with Scheler. He interprets the realm of existing life in Aristotelian fashion as an "ascending scale" but without the evolutionary optimism of a Teilhard de Chardin. Mind, Jonas believes, is prefigured in organic existence from the beginning and manifests itself most prominently in man.

> The great contradictions which man discovers in himself—freedom and necessity, autonomy and dependence, self and world, relation and isolation, creativity and mortality—have their rudimentary traces in even the most primitive forms of life, *each* precariously balanced between being and not-being, and each endowed with an internal horizon of 'transcendence'.
>
> (Jonas 2001, xxiii)

In the terms coined by Schelling, *matter* (or nature) *is not primitive, it is productive* (and creative). This understanding of (matter's or) nature's coming into being and growing is central to Jonas's concept of nature. From an *anthropomorphic* stance,[29] beginning with and descending from man as a *natural* being in whom nature testifies to itself, Jonas discerns "nature's purposive labor" (Jonas 1984, 82) and sees an "immanence of purpose in nature" (ibid., 78), though in a weak sense: Purposes are manifest in nature insofar as organic life carries an ontological "'Yes' of life" in itself: a goal or an intention to continue living, namely the preservation of being. Jonas goes further than Kant's *As-If* and, like Schelling, accuses the mechanists and reductionists of viewing nature *as if* no intrinsic ends and purposes existed. Nevertheless, demonstrating that ends are rooted in nature is not equivalent to saying that conscious intentions and subjectivity are prerequisites of natural purposiveness.

Second, Jonas asserts a *natural purpose as value thesis* or, stronger, an *end as good-in-itself thesis*, namely that "nature, by entertaining ends, or having aims [...] also posits values" (Jonas 1984, 79). Purposes as such, regardless of their material content, represent a value without "judgment upon the goodness of the goal itself" being passed (ibid., 79). The material content of an end or a purpose may be good or bad; ends must initially be regarded as being *good* simply because they exist. Jonas even goes so far as to determine a "purposiveness as good-in-itself" (ibid., 80). He presupposes that there is such a thing as a good-in-itself and that this is not merely a contingent ascription. The good-in-itself is filled with material content—*not* with love, truth, or justice but with purpose. Purposes are posited not only in relative or subjective terms as ascriptions of value (subjective status) but in a more comprehensive way: values "as such" are embedded in purpose (objective or ontological status).

Third, the presence of a good-in-itself says nothing about its actualization. It could be self-sufficient and stand on its own. Jonas, however—with his *good-in-itself as postulation thesis*—attaches to the good an immanent claim or demand: to its being conveyed from potentiality into reality. "[T]he good or valuable," Jonas observes, "when it is this of itself and not just by grace of someone's desiring, needing, or choosing, is by its very concept a thing whose being possible entails the demand for its being or becoming actual" (ibid., 79). It carries an "ought" within itself.

Fourth, even if purposes in nature are bound to an "ought," this does not mean that human beings are able to perceive the "ought" as an obligation and, going further, that they must always fulfil it. The relevant fourth premise for this question differs from the foregoing three. It goes beyond the naturalism of which Jonas was erroneously accused. It can be called a "capacity for freedom" or "capacity for responsibility" argument. The notion of freedom as used here is a *terminus technicus*, which deviates somewhat from everyday word usage. What Jonas means is the active self-preservation witnessed in nature, descending from humans (at the tip of the pyramid) down to the basic forms of organic existence.[30] Humankind, and specifically their freedom, can be regarded as "the supreme outcome of nature's purposive labor," where nature is understood in an integral-monistic sense (Jonas 1984, 82). Jonas's argument is that the special type of freedom inherent in man brings with it an obligation: "This blindly self-enacting 'yes' gains obligating force in the seeing freedom of man" (ibid., 82). It is man's duty to "adopt the 'yes' into his will and impose the 'no' to not-being on his power" (ibid., 82). Recognizing this and acting accordingly are precisely what distinguish *Homo sapiens* and what characterize the human condition. Humankind has to extend its will into an obligation in favour of being, as an "ought-to-be" (ibid., 46f).

To summarize, four premises underlie Jonas's argument of "responsibility for the future," which can be formally expressed as follows: purposive existence + purpose-as-value + good-in-itself + capacity for responsibility ⇒ obligation. Natural purposiveness, Jonas concludes, has a "claim to realization," the willing of which man has the duty to adopt into his "ought." This constitutes the "obligating force of the ontological 'Yes' upon man" (Jonas 1984, 82).

The conclusion is clearly well derived from the argument. However, one may question the premises or theses. Are they true and justified?

First, the question of whether purposes or ends exist in nature has always accompanied philosophy of nature. Despite the Kantian *As-If* devaluation of purposes as a non-objective phenomenon of mere perception and later the positivist and neopositivist verdict against teleology, an absence of purpose is hardly conceivable. Modern biology cannot do without purposes when it gives substance to the various functions of an organism: In this field, the notion of "teleonomy" is quite common. In physical cosmology, the teleologically oriented *Anthropic Principle* describes the selection of initial and boundary conditions realized in the Big Bang and in the early phase of the universe (Carter 1974). In addition, the current advancements in physics, such as physics of complex systems, chaos theory, and self-organization theories, have revived questions of a natural teleology and contributed to a modified, late-modern understanding of nature.[31] Moreover, purposes are articulated in the actions of humans, which can be viewed as part of their nature, following Jonas's or evolution theory. Whenever humans appear so too do ends and purposes. In addition, teleological thinking is part of the current bio-ethical debate, namely in the purview of the potentiality concept: An embryo has the right to develop into a complete human being.[32] Thus, it will not be that easy to dismiss Jonas's natural purposiveness thesis.

Second and third, the "natural purpose as value" thesis and "good-in-itself as postulation" thesis are more problematic. They entail a leap across the *is–ought* or *fact–value* divide. Jonas seems to transgress the Hume–Moore verdict and thereby to commit a categorical or naturalistic fallacy. He, too, does not wish to circumvent this. However, he identifies a tautology in the interdiction, based on an "ontological dogma": Criticism of metaphysics is itself based on metaphysics (Jonas 1984, 131).[33] The alleged cathartic "calling to order"—as Edmund Husserl has named it—is metaphysically charged for its part. It is assumed that "the [concept] serving as the premise here (ultimately borrowed from the natural sciences) is the true and complete concept of being" (Jonas 1984, 44). According to Jonas, the divide presupposes a concept of being (and of nature, man, body, and matter) "that has been suitably neutralized beforehand (as 'value-free')," in which case "the non-derivability of an 'ought' from it follows tautologically" (ibid., 44). The so-called fallacy is then a meaningless proposition; it is not an argument.[34]

The same holds for a further, related dogma pertaining to knowledge or truth. The customary "denial of metaphysical truth presuppose[s] a definite concept of knowledge, for which it is indeed true," namely for the classical, natural-scientific knowledge (Jonas 1984, 44). Jonas's criticism also underpins his own premise that a good-in-itself exists and is manifest in natural purposes beyond any ascriptions.[35] Jonas recognizes that this is not an adequate argument but merely an equivalent point of departure to that of his critics.

Metaphysical elements therefore underlie not only Jonas's ethics but ethical concepts in general. "[I]n every other ethic as well, in the most utilitarian, [...] most this-worldly," he writes, "a tacit metaphysics is imbedded" (Jonas 1984, 44). For example, one might call to mind anthropological, action-theoretical, or physical/material assumptions, which often exhibit a systemic character: It would be rash to want to call them "non-metaphysical." Metaphysics and methodology cannot be totally separated. Accordingly, the objection to metaphysical considerations is not one that applies to Jonas in particular but to ethical conceptions in general. It may encompass everything, but it says nothing.

Such analytic clarification with regard to the inevitableness of metaphysical elements—also highlighting the fact that the critics are not critical enough of themselves—establishes the same initial argumentative conditions for both Jonas and his critics. Jonas goes further and offers proof that bridges between "is" and "ought" do, in fact, exist. He takes a quintessential situation which is paradigmatic for his concept of ethics: a new-born, the "archetype of responsibility." For Jonas, the care relationship between parents and child illustrates the very core of his ethics of responsibility: It involves a "*caring-for* structure," incorporating a "role of stewardship" (Jonas 1984, 8). The "is" of a helpless new-born addresses an "ought" to the world around it (i.e., to take care of it); the new-born articulates an "ought-to-be" and with that a moral obligation: "Look and you know" (Jonas 1984, 131). Even before Jonas, William K. Frankena (1973) had spoken in a similar intuitionistic way of a "principle of beneficence."[36] William James (1977, 619) writes: "Take any demand, however slight, which any creature, however weak, may make. Ought it not, for its own sole sake, to be satisfied? If not, prove why not." In line with Jonas's approach, the direction of argumentation is here reversed: The non-moral is now liable to justification.

Jonas is well aware that one can evade an infant's demands. One can choose a different approach and view the infant in a different way: The infant's "call can meet with deafness" (Jonas 1984, 131). Perception through the senses, inner relating, and participation in the life process of the other are necessary conditions for acknowledging an infant's claims. There is no universal way. What matters is the mode of approach. For instance, one could maintain that a new-born is composed of a conglomeration of cells. Normally, however, when we speak of a "new-born" we are referring to a "being" that, in its existence as such, escapes the reductive approach of classic-modern natural sciences.

In spite of the emphasis on approach, Jonas's argument is decidedly anti-constructivist. It is not we who construct or constitute "being": "being" is granted to itself and to us. Consequently, the appellative function of "being" is precisely not based on a mere ascription. An infant is not merely a physical entity but a subject whose mere being addresses a claim to its environment and in doing so presents itself to its contemporaries as an infant.

Fourth, the premise that man is *de facto* capable of responsibility is considered a standard in moral philosophy: Freedom—including being able to choose between alternatives—is widely taken for granted. Furthermore, it is a position that can also be grounded in dualistic and pluralistic as well as neutral-monistic ontologies. The possibility of responsibility is an obvious necessity for any kind of ethics.

Practice and action—and against the fourth objection

The ethics proposed by Jonas encompasses diagnosis, analysis of causes, and an argumentation framework, but its core lies in praxis: in what contributes to the survival of humans in the technological age. Ethics claims to be good for something. Formal processes of argumentation are necessary but not sufficient.[37] What matters most is the practice. Ethics serves to improve praxis; it *is* praxis in the original sense. The context of practice of Jonas's ethics centres on both the non-reciprocity dimension and the caution or prudence dimension.

First of all, "nonreciprocity of the duties to the future" (Jonas 1984, 38) constitutes the formal core of the necessary condition for the possibility of responsibility: "Responsibility" becomes the central concept.[38] Responsibility as such can ensue only where an asymmetrical structure of power relationships is present: Man today possesses power over future man—but the reverse is not the case. He determines the conditions of existence for humans living in the future. Likewise, humans hold power over animals, adults over children, those already born over those just conceived, fully conscious individuals over the brain-dead, and humans over the natural environment.

Reciprocity, on the other hand—as upheld, for instance, in theories of justice or in discourse ethics—presupposes a symmetrical relationship between free and equal discourse participants: The duty *of one* is the counterpart to the right of *another* and vice versa.[39] Ethics then belongs essentially just to the sphere of social interaction, communication, and discourse among discursively empowered, autonomous subjects. It is assumed that every person is in principle capable of speaking for themselves in the sphere of discourse: It appears uncertain how that is true for future generations, animals, people with dementia, or embryos. According to Jonas, this "scheme [of reciprocity] fails for our purpose [of a global ethics for the future]. For only that has a claim that makes claims—for which it must first of all exist. [...] The nonexistent makes no demands and can therefore not suffer violation of its rights" (Jonas 1984, 38f). The reference to "contemporaneity and immediacy" is insufficient (ibid., 17). The key to an expanded and supplemented ethics, Jonas argues, is its relation to the future: It should also embrace the "not-yet-existent" (ibid., 39). Jonas's aim is to promote a change in the direction and practice of knowledge production, in the course of which scientists acquire an awareness of the ambivalence of their actions in light of their asymmetrical power. For Jonas, such a new mindset implies a professional ethical responsibility of the natural scientist for peace; of the engineer for safety, security, and the environment; or of the physician for the patient. Jonas rejects the molecular-genetic instrumentalization of the "biological material" of life at its beginning and end.

Non-reciprocity can be described as a responsibility-*for* (caring-for configuration or stewardship arrangement) rather than primarily as a responsibility-*to* in the sense of justification vis-à-vis an external instance, an institution, or a person. Accordingly, it would not do justice to Jonas's approach to comprehend responsibility in formal terms as an n-dimensional relation. A formal approach of this kind would void the notion of responsibility of "material" content—namely the duty to ensure the future generations of humankind.

Second, the non-reciprocity dimension provides a general structure to Jonas's ethics—but no more. Where ethical practice is concerned, it needs to be methodologically underpinned and expanded. Jonas introduces a further dimension in this respect: the caution (prudence or tutiorist) dimension. Interlaced with the precautionary principle, the caution dimension forms the methodological core of his catastrophe prevention ethics. Although the caution dimension cannot be inferred logically in a strict sense, it is partly contained within the non-reciprocity dimension: Prudent action taken today refers non-reciprocally to something in the future and therefore can be characterized as responsibility. "[C]aution is [...] a command of responsibility" because "[n]ever must the existence or the essence of man as a whole be made a stake in the hazards of action" (Jonas 1984, 191/37). From a methodological angle, Jonas calls for a "heuristics of fear" with "prevalence of the bad over the good prognosis" (ibid., 26/31). The anti-utopian precautionary principle—with its recognition of an *objective indeterminacy* of real futures[40]—constitutes a conservativism appreciating the "responsibility for existence." Man's biological and socio-cultural existence in the future should be ensured and made possible by laying emphasis on conservation and preservation.

The precautionary principle thus is based on the observation that the future is largely a closed book even if certain anticipatory possibilities are available: In light of the cumulative effects of technological and scientific progress, the future is only partly accessible. Intentional planning that presupposes the possibility of acting according to means/ends schemata is limited. An "essential inadequacy" and "uncertainty of prognostications" have to be taken into consideration when enabling adequate societal decisions on future technologies (Jonas 1984, 31/28). These limitations in terms of knowledge of the future entail that "the prophecy of doom is to be given greater heed than the prophecy of bliss" (ibid., 29/31). Waiver of action will sometimes be necessary, wherein Jonas assumes an asymmetrical relationship between action and waiver of action (e.g., in biotechnology).

Jonas combines both dimensions into a "new imperative" supplementing Kant's formulation with a future-oriented and hence non-reciprocal dimension: "'Act so that the effects of your action are compatible with the permanence of genuine human life'; or expressed negatively: 'Act so that the effects of your action are not destructive of the future possibility of such life'" (ibid., 11).[41] These words articulate a twofold claim to both survival *and* life: the survival of man as a species *and* of human life in the socio-cultural context. In formal terms, Jonas looks to Kant for guidance and takes into account the unconditional obligatory quality of duty with its generalizability. But unlike Kant, Jonas fills his imperative with material content referring to the future possibility of being able to live—and in this requirement, a "sense of morality" must come to the rescue. Moreover, Jonas's imperative, unlike Kant's, possesses little internally logical stringency. Jonas himself concedes that "no rational [= logical] contradiction is involved in the violation of this kind of imperative" (Jonas 1984, 11). After all, it may still be feasible to will the present good while sacrificing the future good.

The imperative of responsibility might sound anthropocentric at first. It is addressed to human beings as acting subjects; it refers not to nature in general but to human beings and to humanity in particular: Jonas postulates a "permanence of genuine *human* life." Yet classifying the imperative as anthropocentric would miss Jonas's intention, as would describing it as holistic. Jonas seeks to circumvent outer ascriptions of this kind by arguing for the common denominator of philosophy of nature, anthropology, and ethics. He sees man from an anthropological perspective, as part of nature conceived of in an integral monistic sense: Man participates in the natural process of life in its entirety. This process also encompasses the socio-cultural context and along with it man's freedom to act (as nature) within nature. It therefore can be said that man is the addressee of the ethical imperative.

Contrary to the view held by some of his critics, the two dimensions of Jonas's ethics—the non-reciprocity and the caution dimension—have not been without impact. They have even found their way into the wording of legislation, as illustrated by examples such as the non-reciprocity dimension incorporated in the German Embryo Protection Act 1990 and the German and Swiss Animal Protection Acts ("animal as living creature" and "dignity of the creature") or the caution dimension found in the *Precautionary Principle* and risk analysis criteria in the context of *Technology Assessment* methodologies.[42] Some years ago, Jean-Pierre Dupuy also evoked the precautionary principle as a metaphysically-methodologically well-founded framework of a new ontology of time for the assessment of nanotechnology; Dupuy (2004, 30) speaks of the necessity of "enlightened doomsaying." The Action Group on Erosion, Technology and Concentration (2007), citing Jonas, demanded a moratorium for nanotechnology.[43]

Jonas has also influenced political programs, such as the Brundtland Commission for Environment and Development, 1987 (WCED 1987). The Commission that propagated the concept of sustainable development was significantly inspired by Jonas. In its concluding report, sustainable development is defined as "a development that meets the needs of the present, without compromising the ability of future generations to meet their own needs" (ibid., 55). Jonas's ethical imperative of responsibility is identical in content even down to the wording[44] and also in terms of the diagnosis of the recent situation of a "threatened future" (WCED 1987, 39). In order to foster sustainability as a political leitmotif, Jonas saw the need for "a new science [...] which [would] have to deal with an enormous complexity of interdependencies," an interdisciplinary, integral "environmental science," which he restated in his book *Technology, Medicine, and Ethics* published in the 1980s (Jonas 1987, 11).[45] Present-day sustainability science, socio-ecological research, and technology assessment can be regarded as concepts of new interdisciplinary sciences of which Jonas spoke. Certain changes have clearly taken place within the sciences and the research system since Jonas's day—without ethically responsible research and development having become the mainstream. Moreover, Jonas had in mind a different type of natural or ecological science encompassing more participatory elements; this position has certainly remained current. In fact, a scientific approach of that type has actually emerged in physics and beyond.[46]

Given all of the foregoing, it seems unwarranted to maintain that Jonas's ethics of responsibility is irrelevant for a problem-oriented approach or for societal praxis.

Prospects: shaping "metaphysics," building society

Jonas is a forerunner of the idea of interdisciplinarity and of new interdisciplinary sciences. His approach is clearly paradigmatic for a critical-reflexive kind of problem-oriented interdisciplinarity: It starts from a shock in the face of the present-day threat. "Modern technology, informed by an ever-deeper penetration of nature and propelled by the forces of market and politics, has enhanced human power beyond anything known or even dreamed of before. It is the power over matter, over life on earth, and over man himself" (Jonas 1984, ix). The inherent ambivalence of scientific and technological advancement "has begun to show its face"—questioning the Baconian project of the modern age in general and, accordingly, challenging ethics and also any domain-restricted philosophy. According to Jonas and his anti-visionary critical-reflexive thinking, despair does not offer a way out. Instead, a different way of framing, perceiving, and understanding nature—which entails a modified mindset—is required to change the human manner of acting within (and the societal relations to) nature. Such a deeper perception of the cultural basis is urgently needed in order to open avenues for an engaged stewardship.

Jonas not only deserves credit historically for putting pressing issues of the life-world, induced by the ambivalent advancement of modern science and technology, on the agenda of philosophy and effectively influencing public debates (cp. Schmidt 2013). He also delivers a strong argumentation for "future-oriented responsibility," grounded in a diagnosis of the current state of affairs and analysis of the reasons for the failure of ethics. Arguably, his metaphysical grounding is the weakest part of his environmental philosophy, as Jonas himself concedes: "I know that this is no proof and compels no one to agreement" (Jonas 1996, 182). "Logically," in the narrower (i.e., formal analytical) sense, an ultimate grounding is lacking (Jonas 1984, 11). On the other hand, Jonas finds the formal rationality of mainstream analytical or transcendental-pragmatic ethics not only foreign to his way of thinking but also inadequate because it is a rationality too closely interwoven with the present crisis: It has even contributed to causing the crisis. Consequently, ethics should always go beyond such an intra-philosophical context of justification and extend—quite in the classical sense—to praxis.

Fundamental issues require fundamental analyses, Jonas points out. For Jonas, the environmental crisis is not a shallow (or surface) phenomenon that one could master solely with modified rules, regulations, or methods—it is a deeper-seated problem rooted in the metaphysical foundations of contemporary culture. As a consequence, reflection on and (potentially) a revision of the underlying concepts regarding the framing, conceiving, and understanding of nature and the societal relations to nature, including inherent metaphysical convictions, are inevitable. This is because environmental ethics is always based on concepts of nature or on what is now known as philosophy of nature, which Jonas calls "metaphysics": "Metaphysics must underpin ethics" (ibid., x). In fact, ethics forms a major part of what Jonas considers "philosophy of nature": "A philosophy of nature is to bridge the alleged chasm between scientifically ascertainable 'is' and morally binding 'ought'" (ibid., x). Most kinds of ethics, and of problem-oriented interdisciplinarity, barely acknowledge underlying metaphysical or ontological elements. They are blind to metaphysics, even if they themselves inherently incorporate a substantial portion of metaphysical assumptions. Jonas considers such kinds to be not only uncritical with regard to themselves but—from a pragmatic perspective—also tame and ineffective and purely cosmetic compensation. They come into play far too late and in a too tentative way. These ethical concepts, he argues, overlook the fact that the alleged "post-metaphysical age" is interwoven with various types of metaphysics.

In sum, the societally relevant question is not whether we have, or want, metaphysics, but which type of metaphysics can and should that be.[47] From this perspective, Jonas can even be read, in the same vein as the scholars of the Frankfurt School, as a critical-reflexive theoretician seeking to engender through sound reasoning—based on his critique of the criticism of the so-called naturalistic fallacy—an enlightened and rationally justified metaphysics which is adequate for dealing with the crisis. Jonas's position could be characterized in the style of the famous work *Shaping Technology / Building Society* by Wiebe E. Bijker and John Law (1994) as *shaping concepts of nature* ("shaping metaphysics"), *changing ways of conceiving nature, building society*. That is to say, the way we think and frame nature and our societal relations to nature influences our decisions and actions. The specialty of a critical-reflexive type of interdisciplinarity is that it considers ontology or metaphysics from a somewhat pragmatic perspective: What kind of ontological background and metaphysical fundament does society at large need to prevent the continuous emergence of new pressing problems? What kind of metaphysics could enable us to change the way we act and make our life more sustainable? A practically relevant environmental *Philosophy of Interdisciplinarity* in which ethics, anthropology, metaphysics, philosophy of nature, philosophy of science, as well as politics and the life-world are conceptualized as a converging domain in a critical-reflexive fashion is proposed here in alignment with Jonas's argumentation. In this chapter, we have seen that such a problem-oriented, ethically reflexive, pragmatic shaping of metaphysics—as a foundational prerequisite for an environmental ethics pertaining to an anticipative shaping of science, engineering, and technology—hardly correlates with our instrumental late-modern self-stylization, which is also predominant in most types of interdisciplinarity. Nevertheless, Jonas has probably hit a nerve, reaching much further than intra-philosophical debates and providing the key to a critical analysis of the metaphysics underlying contemporary society.

What can we learn from Jonas in terms of reflexive philosophical praxis? Philosophy—and, specifically, *Philosophy of Interdisciplinarity*—should not be apathetic or indifferent about the world. It is not a value-free enterprise but rather concerns the world's state of affairs—especially environmental issues and global change problems. It is engaged in changing the situation, for example, by fostering people's awareness, the responsibility of scientists or, in general, humans' stewardship for nature. Such philosophic engagement provides a reflexive fundament for the betterment of societal praxis—and for a good life. It ties in with the original idea of interdisciplinarity as an environmental concept that emerged because of the need to re-orientate technoscientific and societal development: Responsibility and accountability—ethical aspects incorporating a societal stewardship relation of humankind to nature—ought to become embodied in the day-to-day good practice of scientists. To accomplish such a transition, the development of a different mindset towards nature, which will govern our approach to the natural environment and change our societal relations to nature, is required. Jonas is a precursor of a critical-reflexive position in problem-oriented interdisciplinarity.

Notes

1 This book agrees with Frodeman's fundamental analysis (cp. Frodeman 2014).
2 The most comprehensive work on Jonas is by Böhler et al. (2009ff). Introductions to Jonas's work have been presented by Böhler (1994), Levy (2002), Morris (2013), Schmidt (2007a), Schmidt (2013), and Wetz (1994) amongst others. Jonas's predecessors include Leopold (1949), Schweitzer (1966), Picht (1969), and Schulz (1972). In particular, the works of Leopold (1949) and Passmore (1974) are seen as landmarks in the US (bio-)conservation movement (cp. Norton 2005). A cognate approach is pursued by Taylor (1986) and Rolston (1988).
3 Jonas can also be considered the major opponent to the rhetoric of innovation and to Bacon's program—in particular to the Baconian view of nature and the instrumentalist understanding of knowledge—and hence to the instrumentalist concept of interdisciplinarity.
4 First published in 1979 in German.
5 For concepts of socioecological transformations, see Becker and Jahn (2006) and Schneidewind and Singer-Brodowski (2013).
6 In this compendium, the papers are published in German only (Jonas 1987, 11f).
7 The "Anthropocene," as we call the new epoch today, has been emerging.
8 See also Jonas (2009, 126).
9 This motto refers implicitly to a slogan broadly present in the field of science and technology studies (STS) (Hackett et al. 2008): "Shaping Technology, Building Society" (Bijker and Law 1994).
10 See previous works: Schmidt (2007a) and Schmidt (2013).
11 See the usage of the term "philosophy of nature"—and not "natural philosophy"—by Jonas (2001, 282).
12 My translation (J.C.S.).
13 My translation (J.C.S.); see also Krebs (1999).
14 My translation (J.C.S.).
15 My translation (J.C.S.).
16 My translation (J.C.S.).

17 My translation (J.C.S.).

18 For a more detailed discussion of the precautionary principle, see later in this chapter. An in-depth discussion of the precautionary principle can be found in Manson (2002).

19 Others support Jonas's approach and underpin his argumentation, for instance, Böhme (1999) and Dupuy (2004, 2009). In addition, Habermas (2002, 84f.) alludes to Jonas when criticizing the "instrumentalist view of men induced by gene technology." For a deeper reflection on Jonas's work, see Böhler (1994) and publications of the scholars of the Hans Jonas Center, University of Siegen, Germany.

20 This is a translation from the foreword of the original German publication (1979, 7) of the *Imperative of Responsibility*, which was not translated and included in the English version. The overall translation was done by L. Ferry, *The New Ecological Order*, Chicago 1995, 76–77. I have modified it slightly, based on the German original. In the English translation, Jonas writes that "an ever-deeper penetration of nature" is taking place: "[T]he other side of the triumphal advance has begun to show its face, disturbing the euphoria of success with threats that are as novel as its welcomed fruits. [...] The net total of these threats is the overtaxing of nature, environmental and (perhaps) human as well" (Jonas 1984, ix).

21 The latter is my translation from the foreword of the original German publication (Jonas 1979, 7).

22 Jonas does not see his approach as a contradiction of well-established concepts of ethics. Jonas takes from Kant his deontological background, from utilitarianism the consequentialism (including the focus on the results of actions), and from Aristotle the notion of virtues.

23 On the relevance of metaphysics and ontology in Jonas's work, see Hösle (1994).

24 See also Jonas (2009, 126). In his "philosophical biology," Jonas (2001, 282) argues that, "through the continuity of mind with organism and of organism with nature, ethics becomes part of the philosophy of nature."

25 Making metaphysics explicit is an excellent starting point for rational argumentation and critique. Jonas (1994, 45) argues: "[T]he worldly philosopher struggling for an ethics must first of all hypothetically allow the possibility of a rational metaphysics, despite Kant's contrary verdict, if the rational is not preemptively determined by the standards of positive science."

26 My translation (J.C.S.).

27 Most interestingly, Habermas (2002, 86) has shown that the Critical Theory of the Frankfurt School and Jonas's philosophy share a common denominator. According to Habermas, Jonas starts his diagnosis by underlining that techno-scientific development is part of the dialectic of enlightenment, in which man's technical power to rule over nature brings mankind deeper into the internal dynamics and forces of nature: There is no escaping from the roots of mankind in nature.

28 Jonas disagrees that the sciences present the problems that have to be solved by ethics. Such an instrumentalist view of ethics is a shortcoming that does not reach the internal dynamics of sciences and technology.

29 We need to draw a distinction between Jonas's anthropo*morphic* interpretation of organism and life, on the one hand, and his anti-anthropo*centric* (and anti-dualist) approach in ethics, on the other hand.

30 This idea comes close to the descending ontology of Jonas's teacher, Martin Heidegger: What nature is, is not disclosed in molecules, atoms or in particles, but in the most complex structure (e.g., in humans).

31 See, for instance, Schmidt (2007a).

32 In addition, Jonas (2001) argues that the rejection of teleology is not a result of science, but rather an (a priori) assumption.

33   The methodology of natural sciences, as encountered for instance in the guiding idea of Ockham's razor, is based on—mostly implicit—metaphysics (i.e., on a specific preconception with regard to what nature is). This holds also for philosophical positions such as empiricism, positivism, and critical rationalism.

34   Alluding to Jonas and others, Briggle (2010, 93/96f) speaks of "the fallacious naturalistic fallacy."

35   Jonas argues in a programmatic way: "A philosophy of nature is to bridge the alleged chasm between scientifically ascertainable 'is' and morally binding 'ought'" (Jonas 1984, x).

36   Frankena (1973) assumes that the principle of beneficence encompasses various elements, such as the principle of non-maleficence, which states: One ought not to inflict evil or harm; one ought to prevent evil or harm; one ought to eliminate evil or harm; one ought to do or promote good.

37   For instance, discourse ethics (Apel, Habermas) focuses on formal rationality principles, such as procedures of justification and argumentation. Going beyond discourse ethics and formal procedures, Jonas aims to contribute material content and not simply formal elements to ethics.

38   We do, in fact, find a widespread adoption of the notion of "responsibility," accompanied by a decreasing usage of notions like "duty" (Weischedel 1958; Schulz 1972, 632f).

39   For a comparison of discourse ethics and Jonas's approach, see Kuhlmann (1994) and Ott (1993).

40   With reference to Jonas, Dupuy (2004) pursues a similar approach.

41   This wording has evoked a debate on whether Jonas's concept should be seen as a kind of deontological ethics (Grunwald 1996, 196) or, on the contrary, as a consequentialist form of ethics (Lenk and Ropohl 1987, 10/14). Jonas aims at a middle way solution by combining these two approaches.

42   See, for instance, Grunwald (2002) and Liebert and Schmidt (2010).

43   Jonas's approach can be extended and applied to accelerated development in synthetic biology (see final chapter of this book).

44   This observation is more obvious in the German original of *The Imperative of Responsibility*. Jonas uses the term "Dauergebot," which is cognate or derivative to "sustainability" (Jonas 1979, 337); this term is not found in the English book.

45   Published in German only.

46   See next chapter and Schmidt (2008a, 2008c, 2011a).

47   This means that we have to be open to the idea that metaphysics is not unidirectional and determined by the objects, entities, and structures of the world, but also constructed by men (cp. Hampe 2006). We need to consider that we have a certain degree of freedom to construct an appropriate metaphysical fundament—this is consistent with some lines of the pragmatist tradition.

7 Nature and the sciences

In search of alternative concepts of nature and science

Self-organizing phenomena

Let us recapitulate some elements of critical-reflexive interdisciplinarity.[1] This kind of interdisciplinarity is in line with the classic ideal of interdisciplinarity formulated in the early 1970s. It is intertwined with the perception of and reflection on the limits of scientific disciplines or, more generally, the insufficiencies of disciplinary science and their institutionalization in specialized silos. Such critical reflection is far from being an end in itself. Rather, it opens avenues for exploring new directions within the sciences and for fostering a change in the way sciences conceptualize (*ex ante* and *ex post*) nature and our societal relations to nature. In addition, it can encourage scientists (and all of us) to question what counts as legitimate science, entailing a cultural critique of present-day fragmented knowledge production, the institutionalized research system, and the related (Cartesian dualistic) worldviews.

Critical-reflexive interdisciplinarity is interlaced with a transformation of our scientific mindsets. It thus can be associated with self-reflexivity, self-awareness, and self-critique or, to put it briefly and more provocatively, with self-enlightenment—specifically with considering and facing our own blind spots, our non-knowledge, and our ignorance. Critical reflection on the limits of one's own approach to the world and on the boundaries of a particular framing of the world's objects not only serves as a prerequisite or initial point of departure but also is an indispensable part of any reflective interdisciplinary practice, at least in connection with what we have labelled critical-reflexive interdisciplinarity. Because self-reflexivity is not to be understood as an ultimate state that can be reached, maintained, and preserved but as an ongoing process and non-finishable challenge—being a core element of what Erich Jantsch (1972) once called "a continuous process of self-renewal"—critical-reflexive interdisciplinary practice is linked to specific (i.e., interdisciplinary) virtues, mindsets, and habits.[2] This kind of interdisciplinary practice goes beyond the definition of a common learning process or of stakeholder involvement; it is a mode of continuous (self-) critique and (self-)reflection.

DOI: 10.4324/9781315387109-8

In addition to this *first* aspect of critical-reflexive interdisciplinarity, three further dimensions deserve attention in the endeavour to do justice to the concept of critical reflexivity. *Second*, critical-reflexive interdisciplinarity aims to achieve a synthesis or synopsis. A synthetic-synoptic perspective—meaning the development of a coherent yet at the same time non-reductionist way of seeing things—complements the critical orientation as a second fundamental constituent of critical-reflexive interdisciplinarity. The goal of critical-reflexive interdisciplinarity is to bring together knowledge and insight from different domains and disciplines or sometimes also from the non-disciplinary public—in order to build and shape an alternative conceptualization of nature and of the societal relation between humans and nature.[3] *Third*, critical-reflexive interdisciplinarity also aims to promote a change or transformation in the orientation of science and scientific advancement. It looks for inner-scientific alternatives beyond mainstream science. The final and *fourth* aspect, as we will recall, is that critical-reflexive interdisciplinarity is subsumed under the umbrella of problem orientation since it faces or relates to grand societal challenges. More specifically, it addresses the background and underlying causes of these challenges and in particular the societal relations to nature.

As discussed in the preceding chapter, none other than Hans Jonas pursues these four directions of critical-reflexive interdisciplinarity—(1) critique (of, towards, and within the sciences), (2) synthesis (in order to advance a different view of nature and of the human/societal relations to nature), (3) new directions and alternatives (within the sciences), and (4) a certain kind of problem orientation—in suggesting that "it is rather the evolutionary view of nature as becoming than the stable and enduring one [...] that provides prospects" in this regard. Jonas (1997, 402/18)[4] draws on the "potential of self-organization of matter towards life" and, by doing so, argues for a "deep revision of how we conceive and conceptualize nature" as the ontological or metaphysical fundament of his *Imperative of Responsibility*. Jonas's view of nature is in line with Erich Jantsch's opus magnum *The Self-Organizing Universe* (Jantsch 1980). Jantsch argues for a "self-renewal of science" entailing a participatory and "process-oriented recognition" of the natural, social, and human world that includes the emergence of the human mind and creativity at large (Jantsch 1980, 3/6). Jantsch's vision comes close to the thinking of the Nobel Laureate Ilya Prigogine and the philosopher Isabelle Stengers (1984), who claim that Western thought tradition as manifest in most exact disciplinary sciences—with their worldviews, presupposed ontologies, and their *ex ante* metaphysics—has far too long emphasized static-stable "being" at the expense of dynamical-evolutionary "becoming." Timelessness and temporal invariance were the deceptive metaphysical convictions that entailed an overall framing of the world through the lens of stability; nature was, and to some extent still is, conceived of and conceptualized as nature insofar as it is stable or, taking it further, even static. However, that view of nature—which Plato and later Bacon and Descartes had also put forward and which was substantiated by Newton and Einstein—does not only lack empirical evidence. Even more relevant is that such a mindset brings with it a misleading conceptualization of nature as a dead, non-evolutionary mass or as a mere mechanical assemblage of atomic or sub-atomic entities that appears to be entirely at the disposal of human beings: Why should we take care of nature, Jonas asks, if nature is no more than a dead block that, like a mechanical gear, can be easily repaired? This way of perceiving nature has played a central role in causing the environmental crisis, as Jonas, Prigogine, and Jantsch argue from cognate angles.

In light of self-organizing phenomena, Jonas as well as Prigogine and Jantsch encourage us to develop a new way of looking at nature and, in particular, to change how we perceive and conceptualize nature and our societal relations to nature; they urge us to adopt a synthetic but pluralist and participatory view of the things. Certainly, this seems a relevant perspective for dealing with our socio-ecological crisis insofar as our way of viewing nature is inherently interlaced with how we act and conduct ourselves towards nature. In other words, ontology and ethics are not to be kept strictly separate. Not only are we failing to tackle recent environmental problems; more severely, our incapacity to prevent the emergence of new environmental problems signifies that our framing and conceptualizing of nature are faulty, insufficient, and disastrous, as Jonas argues: The ongoing production of new problems falsifies our ontological concepts and metaphysical convictions of what nature and our societal relations to nature are. In this vein, Heidegger (1977, 40) urges us to rethink our thinking: "Before considering the question that is seemingly always the most immediate one and always the most urgent [ethical] one: 'What shall we do', we ponder this: 'How must we think'?"[5] Therefore, questions concerning our framing, perception, and conceptualization of nature—namely our mindset—are central to a problem-oriented approach. Moreover, an efficient and effective problem-oriented interdisciplinarity that digs down to the roots and considers the underlying causes of the environmental crisis requires an adequate (plural and provisional) ontological conceptualization of nature. A good problem-oriented interdisciplinary practice is critically reflexive.

Such an orientation of the kind demanded here—comprising synthetic (see next section), critical (see section after next), alternative (see further section), and problem-related (see further section) dimensions—is gradually emerging in a certain branch of interdisciplinary open physical sciences. That is the thesis of this chapter, which intends to expound an alternative to the mainstream sciences. It will also be shown that the very idea of what can count as legitimate science is changing. In the process, a modified view of nature seems to be evolving.

Synthesis—a first dimension of critical-reflexive interdisciplinarity

Self-organization theories

Since the advent of self-organization theories and their progress in the realm of physical sciences during the 1960s—facilitated by the advancements in computer technology and the computer's capacity to numerically handle a certain type of mathematical (i.e., nonlinear) model—further evidence of the existence and prevalence of unstable, self-organizational, complex phenomena in different disciplinary domains has emerged. The ways in which nature and science are viewed are changing.

The paradigm of self-organization—substantiated by self-organization theories or, synonymously, by complex systems—represents one of the most striking hallmarks of what can be deemed an interdisciplinary, structural-scientific revolution. Self-organization theories are paradigmatic for structural sciences. The term "structural science" was coined by the physicist and philosopher Carl Friedrich von Weizsäcker (1974, 23). "Structural sciences," Weizsäcker wrote back in the 1950s,

> encompass systems analysis, information theory, cybernetics and game theory. These concepts consider structural properties of various objects regardless of the material realm or disciplinary origin. Time-dependent processes form a common umbrella that can be described by an adequate mathematical approach and by using the powerful tool of computer technology.[6]
>
> (ibid.)

Today, we can add to Weizsäcker's list: self-organization theories—encompassing complex systems theory, nonlinear dynamics, chaos and catastrophe theories, synergetics, fractal geometry, dissipative structures, and the like. Structural sciences focus on the analogies or structural similarities shared among different objects of various disciplines. Bernd-Olaf Küppers (2000) argues that we are experiencing a "structural convergence of physics, chemistry, biology and computer science."[7] Today, a "structural scientific revolution tears down disciplinary boundaries," brings different disciplines together, and provides "explanations on the systems level."[8]

What is the common, discipline-transcending denominator of a structural science such as self-organization theories? Is there any integrative element entailing an interdisciplinary synthesis? Indeed, there is. Instability is the crucial factor. Instabilities are the core element and the source of nature's capacity to enable self-organization. Without instability, there is no change, no emergence, no evolution, and no complexity and thus no self-organization and no dynamical stabilization. In the face of the familiar phenomena around us, instabilities cannot be considered exceptions within a stable world. Rather, it is the other way around: Stability is only local, it is restricted to certain spatial and temporal domains—and it is based mostly on instability on more fundamental scales.

Disregard of instabilities

Given the centrality of instabilities in the world we live in, the disregard of instabilities in the history of science appears strange and surprising.

Traditionally, nature was categorically defined by stability. Stability metaphysics dominated our history. Since ancient Greek times, it was not doubted that instabilities indicate a lack of knowledge. Nature, stability, and knowledge were used as synonyms. Plato's cosmos was structured by a Demiurge according to the idea of simple mathematical laws, intrinsic harmony, time-invariant order, and universal stability. Even if nature's harmony, simplicity, and stability seemed obscure at first glance, there was no doubt about the existence of such characteristics behind the apparent complexity, guaranteeing the timeless and infinite existence of nature as a whole.

Plato's foundation of science was so influential that, from the beginning of modern science in the early 17th century, mathematical invariants—from Platonic bodies to Newton's laws and Einstein's equations—were considered the core or essence of nature. In modern times, stability metaphysics can be found in the works of outstanding physicists such as Newton and Einstein. For instance, in his *Opticks*, Newton did not trust his own nonlinear equations for three- and n-body systems which potentially exhibit unstable solutions (Newton 1983). He required God's frequent supernatural intervention to stabilize the solar system. In the same vein, Einstein (1917) introduced ad hoc, and without any empirical evidence or physical justification, the cosmological constant in the framework of general relativity to guarantee a static and stable cosmos, "Einstein's cosmos." Newton's and Einstein's views of nature illustrate that metaphysical convictions—*what nature is* or *has to be!*—can be incredibly strong, even if they contradict what is known about nature at a particular time. In other words, modern science did not, on the whole, scrutinize whether nature was stable or not. Science operated on the implicit stability assumption: All mathematical laws, models, and theories had to be stable in order to represent a stable world. Indeed, framing nature as "nature" insofar as it is stable was a successful strategy to advance a specific physical knowledge.

A more in-depth look shows that the disregard of instabilities in the history of sciences is even more remarkable since physical scientists have always been aware of the existence of unstable processes in nature and in models. The problem has been broadly identified ever since the development of hydrodynamics in the 19th century. The basic equations in this field include the Navier–Stokes equations describing unstable phenomena such as turbulent flows. In addition, Poincaré's description of the solar system and Einstein's general relativity theory reveal several types of instabilities. Even before, instabilities had been taken into consideration in Newton's theory of the moon and later in Maxwell's classic mechanics in the 19th century: "matter and motion." Maxwell was a particularly prominent precursor in identifying instabilities. According to Maxwell, in physics

> there is [… a stability] maxim which asserts 'That like causes produce like effects'. [But,] [t]his is only true when small variations in the initial circumstances produce only small variations in the final state of the system. [...] [T]here are other cases in which a small initial variation may

produce a very great change in the final state of the system, as when the displacement of the 'points' causes a railway train to run into another instead of keeping its proper course.

(Maxwell 1991, 13f)

Maxwell continues by analysing the methodological challenges in cases of unstable systems: "[O]nly in so far as stability subsists, principles of natural law can be formulated; it thus perhaps puts a limitation on any postulate of physical determinacy such as Laplace was credited with" (Maxwell 1991, 13/footnote).[9] Maxwell was the first to explicitly identify the relevance of sensitive dependence on initial conditions—which is the nucleus of dynamical instability.

Beyond the relatively few examples of pioneering insights, the prevalence of instabilities was not obvious in the 19th century. Instabilities were not broadly acknowledged. Instead, they were regarded as exceptions within a stable world; instability was located just on the fringes of stability.

The recognition of instabilities—the core of self-organizing phenomena

The idea that instabilities exist merely on the fringes of stability changed from the 1960s on. As we have seen earlier, that was the decade in which we experienced the beginning of an interdisciplinary or, more specifically, a structural-scientific revolution caused by the rapid advancement of microelectronics and the development of computer technology. Many ideas were developed simultaneously. Ed Lorenz (1963) discovered instabilities in weather forecast, Hermann Haken (1980) developed his synergetics, and Ilya Prigogine and Paul Glansdorff (1971) formulated the nonlinear thermodynamics of far-from-equilibrium dissipative structures. In the early 1970s, David Ruelle and Floris Takens (1971) presented an influential paper on solutions of the hydro-mechanical Navier–Stokes equations and coined the term "chaos" metaphorically for a certain class of solutions to partial differential equations. In 1975, Tien-Yien Li and James Yorke (1975) published an epoch-making mathematical article on one-dimensional difference equations, highlighting "chaos" explicitly in the title and providing a mathematically clear definition: "Period Three implies Chaos." During the 1970s, Benoit Mandelbrot (1982) elaborated his "fractal geometry" and René Thom (1975) presented the "catastrophe theory." It was an exciting time of change in how nature and sciences were viewed, inducing a *structural-scientific* paradigm shift. The main overarching achievement was that instability was acknowledged as a basic feature of nature, and the relation between stability and instability was reversed. No longer was instability deemed a minor subset of stability. On the contrary, stability became one tiny island in an ocean of instability: "order at the edge of chaos." Instability—not stability—is the prevalent case in nature; such insight into the very fundament of nature should not be regarded as simply a negative message of destruction.

Instabilities indicate situations in which the system is on a razor's edge: criticalities, flip points, thresholds, watersheds, and sharp wedges. They generate butterfly effects or sensitive dependencies, bifurcations, points of structural changes, and phase transitions. The list of examples is extensive: the emergence and onset of a chemical oscillation, the roll dynamics of a fluid in heat transfer, an enzyme kinetic reaction, a gear chattering, or turbulence of a flow. A fluid becomes viscous, ice crystallization occurs, a phase transition from the fluid to a gas phase takes place, a solid state becomes super-fluid, laser light issues, a water tap begins to drip, a bridge crashes down, an earthquake or a tsunami arises, a thermal conduction comes to rest, and a convection sets in (e.g., Bénard instability). New patterns and structures appear. Instabilities are the necessary condition for novelty—and for dynamical stability on a higher level, for metabolism, and for living organisms.

Today, ex post and thanks to the advancement in physical sciences, we can identify a "dogma of stability" that has implicitly determined the selection or construction not only of the models, laws, and theories but also of the objects of study. "We shall question the conventional wisdom that stability is an essential property for models of physical systems [...]. The logic which supports the *stability dogma* is faulty" (Guckenheimer and Holmes 1983, 259). The stability assumption itself is unstable! Our world is essentially a world of emergence, dynamics, change, temporality, and evolution. The mere empirical fact that emergence and, more generally, self-organization are possible in the world we live in provides evidence that instabilities are not a mere convention but actually exist in the world. From an ontological and epistemological perspective, this point gives substance to a position that could be called *minimal instability realism*. In this vein, the physicist J.S. Langer (1980) underlines the role of "instabilities for any kind of pattern formation." According to Werner Ebeling and Reiner Feistel (1994, 46), "self-organization is always induced by instability of the 'old' structure via small fluctuations." "This is why studying instability is of major importance."[10] Gregory Nicolis and Ilya Prigogine (1977, 3f.) argue that "instabilities" are the "necessary condition for self-organization." Wolfgang Krohn and Günter Küppers (1990, 3) emphasize that "instabilities are the driving force for systems evolution and development."[11] In addition to referring to novelty, some scientists in this field refer to the conditions of life or the "arrow of time."[12] Traditional, classic-modern (physical) science was "unable to tackle and to resolve the antagonism between reversibility and irreversibility. To accomplish this, it was necessary to introduce a new concept: the concept of instability" (Ebeling and Feistel 1994, 197).[13]

Characterizing self-organization

Instability is the central condition for the possibility of self-organization and emergence. Although that characterization draws on a fundamental ontological element, it is not very specific. What can be deduced from this crucial fact? Since the notion of "self-organization" refers to the formation of new patterns, the emergence of new order, the creation of new structures, or the becoming of novel entities, it has at least two necessary dimensions: novel order, on the one hand, and the process of creating novelty, on the other. In other words, (1) *novelty* and (2) *processuality* (or *temporality*)— enabled by instabilities—constitute the very core of the notion of self-organization. These two characteristics align with the works of two pioneers in the research on self-organizing phenomena—with Hermann Haken's *synergetics* and Ilya Prigogine's *dissipative structures*.[14] Whereas Haken focuses mainly on new order, structures, and patterns (Haken 1980),[15] Prigogine advances a complementary, processual view of nature and emphasizes the aspect of "becoming" (Prigogine 1980). In addition to the characteristics of "novelty" and "processuality," the notion of "self-organization" points to (3) an internal source of the emergence of new order; no external force or actor is necessary. The *internality* of pattern formation is expressed in the prefix "self" of the word "self-organization."

Besides physical systems, biological systems or living organisms are particularly prominent examples illustrating the overall enabling capacity of instability—and the three associated characteristics of self-organization: novelty, processuality, and internality (Schmidt 2015a). Living organisms are, in general, open systems through which a continuous flow of material, energy, and information takes place. In fact, they are not stable in the sense that they attain steady-state equilibrium. A steady state would mean that they are dead; stability as understood in the physical sciences entails such a situation. However, living organisms do not follow the laws of equilibrium thermodynamics but the principle of nonlinear thermodynamics of irreversible processes far from equilibrium. Being open systems, they manage to maintain their overall structure for a while—that is why we speak of dynamic equilibria or pseudo-steady states. A closer look at the models of the underlying mathematical structure—which are based on nonlinear thermodynamics (cp. Prigogine and Glansdorff 1971; Nicolis and Prigogine 1977)—would reveal that, in order to maintain the overall temporal stability of the whole living organism on the mesolevel, instabilities are a prerequisite for enabling the flow of material, energy, and information through the organism on the microlevel. Therefore, stability and instability are not disjunctive; stability in complex systems such as that in (and of) living organisms has to be regarded as a derivative of instability. Certainly, this insight reverses the view about the relationship between stability and instability.

The foregoing analysis indicates that the task of a *Philosophy of Interdisciplinarity*—in line with the classic perspective of natural philosophy—is to bring together the very different perspectives and to suggest a common ground which is open to revision: Synthesis is indispensable for interdisciplinarity. In the case of self-organization, instabilities provide the common ground that enables a novel synthetic-integrative view. Self-organizing phenomena are based on instabilities; the primacy of instability can also be shown for cognate phenomena such as chance, randomness, chaos, turbulence, fractals, symmetry breaking, phase transitions, time's arrow/ irreversibility, and information loss or gain.[16] Thus, instabilities can be seen as the major constituent of the three ontological characteristics of self-organization: (1) novelty, (2) processuality, and (3) internality ("self"). That is, instabilities are ontologically central for any concept of self-organization.

An ontological definition of this kind is not all that needs to be taken into account when providing sense to self-organization. Self-organization is a much richer and more multifaceted concept. A semantically content-ful understanding of self-organization encompasses further aspects that are epistemological and methodological in nature, as we will see below: (4) limits in reproducibility/repeatability, (5) obstacles to predictability, (6) restrictions in testability, and (7) limits in describability and reductive explainability.[17] Like the first three dimensions (ad 1–3), the epistemological and methodological dimensions of self-organization (ad 4–7) are a consequence of instabilities.[18] Most interestingly, these latter dimensions challenge—and also critique—our underlying assumptions regarding what science is and ought to be, as will be discussed in the next section.

Critique—a second dimension of critical-reflexive interdisciplinarity

Landscape of instabilities

Instabilities are a double-edged sword. On the one hand, their acknowledgement advances a synthetic and novel picture of nature, and it paves the way for a fascinating discipline-transcending endeavour of the sciences; on the other hand, instabilities reveal limitations and obstacles within the sciences, which cannot be overcome by minor readjustments, as will be shown. This inherent "dialectic" is characteristic for the critical-reflexive type of interdisciplinarity.

Let us look at the latter point concerning the limits induced by instabilities. What, specifically, do we mean by instabilities? What are they exactly? In a nutshell, three kinds of instabilities can be distinguished.

First, static instabilities or, synonymously understood, watersheds are ubiquitous in nature and in our life-world. The related phenomena are well known from our everyday experience. For instance, a pen standing on its tip will eventually fall down to one side or another. Similarly, a ball on a roof ridge will roll down to one side or the other. The same occurs with a mass of a physical oscillator at the unstable point of maximum potential energy. In a pinball machine, the ball hits sharp wedges and jumps to the right or left. The quincunx, or "Galton's apparatus," consists of rows of alternating pins that are the source of instabilities. A tiny ball that falls on a pin passes it either on the right or on the left. This occurs with the ball several times on several rows. Ultimately, we observe a random-looking Gaussian distribution of all tiny balls at the bottom. That is why instabilities once served as an (implicit) underlying idea for the development of classic probability theory. For Jacob Bernoulli, instabilities in coin tossing—and the resulting binary sequence of 0 (pitch) and 1 (toss)—are paradigmatic for producing randomness. Instabilities generate nomological randomness or, to put it in other words, deterministic chance.

The concept of static instability can be traced back to the end of the 19th century, when Henri Poincaré inquired into issues of stability and instability within the planet system and, more specifically, within the three-body system sun–earth–moon (Poincaré 1892): "A very small cause which escapes our notice determines a considerable effect that we cannot fail to see, and then we say that the effect is due to chance" (Poincaré 1914, 56f).[19] At points of static instability, a sensitive dependence on initial conditions occurs. Alternative trajectories from two nearby initial points separate and will never again become neighbours. In the course of the system's time evolution, one raindrop falling near the watershed may be transported to the Mediterranean Sea while its neighbour goes to the North Sea. Between the two raindrops, there is a basin boundary or watershed—in other words, points of static instability. A point worth making here is that the fact that small differences in the initial conditions can cause large effects in the future not only is characteristic of what is known as deterministic chaos but is—as we have seen—much more common.

Overall, time evolution is of minor relevance for cases of static instability. After a certain time, the initial point has left the neighbourhood of the watershed; the watershed no longer has any impact on time evolution. In this context, instability is not to be considered an epistemological or methodological challenge.

Let us next look briefly at, *second*, dynamical instability, which is frequently called deterministic chaos. Dynamical instability has a more far-reaching relevance and impact compared with static instability. According to Friedrich Nietzsche (1930, 127), the "underlying character of the world is chaos."[20] For Nietzsche, who was a contemporary of Poincaré, chaos is not associated with lawlessness and disorder but with an unforeseeable dynamic or a complex time evolution. His view of chaos differs from the messy *Tohuwabohu* of the Judeo-Christian tradition. That is why Nietzsche can be regarded as a precursor of our recent understanding of "chaos." Like Nietzsche, Martin Heidegger (1986, 87) advocates a positive understanding of chaos: Chaos means "dynamics, change, transition—whereas regularity seems to be hidden: we do not know the underlying law at first glance."[21] Clearly, Nietzsche and Heidegger did not define "chaos" in scientific terms. It was the task of the mathematicians of the 1970s to provide a distinct definition.

Deterministic chaos is essentially characterized by dynamical instability. The time evolution of a chaotic system is continuously on a razor's edge. Nearly all points in the state space exhibit the property of sensitivity: The trajectories separate exponentially by time evolution. Because chaotic attractors are bounded in the state space, we find exponential separation merely as a mean value. For short time scales, two trajectories may converge before they diverge again and so on—an interplay of divergence and convergence among trajectories occurs. However, dynamically unstable (chaotic) trajectories do not show an exact recurrence. Henri Poincaré, the precursor of modern nonlinear dynamics and self-organization theories, studied the phenomenon of recurrence more than one hundred years ago. As Poincaré observed, the dynamically unstable trajectory of a chaotic attractor resembles a ball of wool. It covers the attractor densely. Nearly all points are reached approximately by the trajectory after a certain time (e.g., in the case of a chaotic double pendulum).

Chaos—and its core: dynamical instability—is associated with random-like behaviour generated by deterministic models in a spatial-continuous state space in which every point is a watershed or divergence point: a random *pheno*type in spite of a nomological *geno*type. Although Poincaré and later George Birkhoff (1927) were aware of the possibility of random-like behaviour, they did not have computer resources and numerical simulation tools to deal approximately with dynamically unstable systems. In the 1960s, Ed Lorenz, who discovered chaos in atmospheric dynamics through computer simulations, coined the term "butterfly effect" to illustrate its real impact: For dynamically unstable systems, a tiny event such as a butterfly's wing flap in South America can cause a thunderstorm in the US.[22] The butterfly effect is frequently associated with "weak causality" in order to underline that dynamical instability can still be considered part of a causal nexus of nature, though not in the traditional sense of "strong causation."

Finally, the *third* type of instability, structural instability, has not received the attention it deserves, although it is central to nature—for enabling biological and cosmological evolution and for securing an overall stability of organisms. Structural instability is associated with small changes in a system's structure. This type of instability occurs even in simple physical systems, such as in the heating of water or the Rayleigh–Bénard convection where a fluid layer is confined between two thermally conducting plates and is heated from below, creating a vertical temperature gradient. For lower temperature gradients and hence for smaller density gradients, an upward conductive heat transfer through horizontally uniform layers takes place. If the temperature gradient is sufficiently large, a point of structural instability or, synonymously, a bifurcation point or criticality occurs: The hot fluid rises, leading to a circular convective flow in three-dimensional cells, which results in enhanced transport of heat between the two plates.

In contrast to what has been discussed in the case of dynamical instability, structural aspects are far more fundamental: Structural instability does not refer solely to initial points or a particular trajectory but to the whole dynamical system, represented by the mathematical structure of the model itself, namely the law or equation. Structural instability is generally expressed as follows: If one alters the structure of a model (equation, law) slightly, the overall dynamics changes qualitatively.[23] The equivalence class of the system or model is not preserved; the disturbed and undisturbed models show different topological characteristics, such as different types of orbits or at least different periodicities. Such a case occurs, for example, in the onset of a chemical oscillation or of fluid roll dynamics under heat transfer, in the emergence of laser light, or in a water tap beginning to drip.

Structural stability (and, conversely, instability) was introduced by the Soviet physicist Aleksandr Andronov and mathematician Lev Pontryagin (1937) with the notion of "robustness" or "coarse-grainedness." Birkhoff (1927) is another forerunner of the concept of structural instability. In his influential book *Dynamical Systems*, he defines and distinguishes between different types of stability. Some of Birkhoff's definitions paved the way and gave rise to what would later be known as structural instability. When *dynamical systems theory* was further developed, Andronov's and Pontryagin's term "robustness" was replaced by "structural stability" with no change in content. Structural stability, with its converse of structural instability, is today regarded as a major part of bifurcation theory (Wiggins 1988).

The advancement of complex self-organizing systems led to the acknowledgement that structurally unstable phenomena in nature are not just exceptions in an ocean of stability; rather, the reverse is the case. According to Devaney (1987, 53), structural instability is most prevalent, even in classical physics. "Most dynamical systems that arise [even] in classical mechanics are not structurally stable." In light of its prevalence, structural instability deepens and extends our discussion. Since Plato's time, theories (or models, laws, and equations) have been considered the nucleus of nature as well as the goal of science. In contrast, initial conditions—most important in cases of static and dynamical instability—were seen as merely contingent factors or an outer appendix. Given that structural instability does not refer to initial conditions but to theories and their law-like structure, it can be regarded as the most relevant type of instability.

Instability as a point of critique—and a challenge to scientific methodology

How central is the recognition of instabilities for our understanding of physical sciences and their methods?

It is undeniable that methods constitute the core of science. Although the debate is still ongoing without a final consensus having been reached, scientific methodology can be said to comprise four central, interlaced elements, which also highlight specific objectives and different understandings of science: (1) reproducibility/repeatability/experimentability/technical controllability, (2) predictability/ calculability, (3) testability/objectivity, and (4) describability/explainability/reducibility.[24]

Instabilities—and self-organizing phenomena insofar as instabilities constitute their ontological core—pose challenges to these methodological elements and reveal limitations. The crucial point is that in the case of instabilities the "devil is in the detail." The systems under consideration are on a "razor's edge." In the following, we will look at instability-based issues in the physical methodology.

One, critique of experimentability and reproducibility

Since Galileo and Bacon introduced the experiment in the late 16th and early 17th century, experimentation has been regarded as the methodological core of modern sciences. By means of intentional interventions in nature, we constitute, construct, and control empirical facts of nature. These facts are mostly not given but have to be produced or created (Hacking 1983). According to Francis Bacon and Immanuel Kant, experiments serve as "interrogation techniques" to "force nature to reveal her secrets." Experiments—and not passive observation with unaided senses—guarantee empirical evidence, objectivity, or at least intersubjectivity: Anyone performing a particular experiment in any place and at any time will obtain the same result—if they act in accordance with the norms of experimentation. Classic-modern science is based on the methodological assumption of person, location, and time independency.

Scientists presuppose that they can, at least in principle, control the experimental setup, including the relevant boundary conditions. In other words, they assume the validity of the *principle of isolation*, namely that there is a clear-cut *system–environment* interface and that we can sufficiently isolate the system from its environment. For example, we do not expect a bumble bee on the double star Sirius or a falling leaf in another country to have any influence on the experiment. Otherwise, a particular phenomenon under consideration would not be reproducible. In its core, the distinguishing character of an experiment is therefore repeatability. Scientists have, for a long time, not doubted that reproducibility or replicability is in principle feasible, even if it may not yet have been accomplished with regard to a specific phenomenon. Jürgen Mittelstraß (1998, 107) believes "reproducibility [...] is the major scientific norm." "The requirement of reproducibility is an indispensable criterion to define 'science' as science: in fact, it is the necessary condition for the possibility of sciences and scientific knowledge!"[25] Robert Batterman (2002, 57) stresses that "any experiment in which a phenomenon is considered to be manifest must be repeatable." Gernot Böhme and Wolfgang van den Daele (1977, 189) argue that the "methodological concept of science is the regular fact that includes the condition under which it can be reproduced by anybody any time."[26] For Friedrich Hund (1972, 274), physical science is the "realm of reproducible phenomena."[27] Wolfgang Pauli (1961, 94) emphasizes that—in spite of astrophysics and cosmology—"exceptional and extraordinary events are far beyond the scope of physics; these events cannot be grasped by experiments."[28]

If we were to subscribe to these statements and restrict science to reproducible phenomena, we would find ourselves in a dead end: Science and instability would be like fire and water. But this is not the case, as already shown and as I will elaborate in detail later on. Admittedly, the challenges induced by instabilities are severe. Instabilities convey unobservable small effects to empirically accessible scales. That is why they bridge different spatial domains (e.g., microcosm, mesocosm, and macrocosm) and induce problems regarding experimentation. The devil is in the detail. Because of thermodynamic and quantum mechanical effects, initial and boundary conditions cannot be measured exactly or be controlled in every detail. Making reference to the three-body problem and its instability, Poincaré (1914, 56) emphasizes that "even if it were the case that the natural laws no longer had any secret for us, we could still only know the initial situation approximately." Inherent inexactness causes severe problems for physical sciences. Repeatability is limited; the intentional reproduction of events is difficult to achieve with unstable objects; the dynamics of unstable objects cannot be controlled by the experimenter. However, the lack of control is not just a pragmatic or epistemic boundary that could be overcome by improved methods and more advanced technology. It is part of nature since it is inherent to objects and not just a challenge to knowledge: It is ontological rather than epistemological or methodological.

Two, critique of predictability and calculability

Predictability has been one of the most important qualifiers of what constitutes science. Among those who embrace predictability are instrumentalists and pragmatists and sometimes others such as realists and transcendental epistemologists. Carl Friedrich von Weizsäcker (1974, 122) maintains that "predicting the future" is the "major essence of exact science."[29] In line with Weizsäcker, Michael Drieschner (2002, 90) emphasizes: "We find 'prediction' as the key term for understanding and defining physics."[30] Einstein, Podolsky, and Rosen (1935, 777f) regard predictability as the sole possibility for ascertaining whether any entity actually exists. "If, without in any way disturbing a system, we can predict [at least in principle] with certainty the value of a physical quantity, then there exists an element of physical reality corresponding to this physical quantity." Wesley C. Salmon's (1989, 119) main criterion for a good explanation is the success of predictions: Explaining means to "show that the event to be explained was to be expected." According to Stathis Psillos (1999, xix), scientific realists regard "predictively successful scientific theories as well-confirmed and approximately true of the world." It is well known that Auguste Comte once said: "Savior pour prévoir" (knowing in order to foresee) (cp. Comte 2006).

Instabilities, however, challenge the prediction-oriented understanding of physical sciences. Mario Bunge (1987, 188) perceives the "immense mathematical difficulties with nonlinear laws" and with unstable solutions.[31] A differential equation—such as Newton's or Einstein's laws of motion—provides a general corpus for a law of nature but not a precise solution or a particular trajectory. The specific solution is not given with the differential equation itself; it has to be computed in order to enable predictions. Linear differential equations pose no trouble; most of them can be integrated analytically. In contrast, nonlinear differential equations often possess no analytic solutions. Newton, Euler, Laplace, Poincaré, and others were frustrated by this fact. Even if we were ever to succeed in finding a nonlinear theory of everything—the equation that governs the universe—we likely would not be able to predict specific events in the far future. According to Maxwell (1873, 440), "it is manifest that the existence of unstable conditions renders impossible the prediction of future events, if our knowledge of the present state is only approximate, and not accurate." In some cases, numerical algorithms can help to integrate nonlinear differential equations and handle unstable solutions.[32] But in other, most prevalent cases of dynamical instability, unstable chaotic orbits cannot be approximated by numerical solutions. They are effectively non-computable. Robert Devaney (1987, 53) stresses that "most dynamical systems that arise in classical mechanics are not stable. [...] These systems cannot be approximated" employing the classic approach of asymptotic convergence of mathematical functions. No numerical solution is accurate enough to determine an unstable trajectory; for unstable systems, these prediction tasks would require an accuracy that is *de facto* impossible (cp. Schmidt 2011a).

There are two kinds of obstacles involved here. A *first* numerical obstacle is due to the impossibility of digitizing real numbers. Digital computers calculate and store results only to finite precision, so there are rounding errors at every calculation step. In an unstable system, these errors will grow exponentially and so the model's trajectory (when initialized from a particular state) will quickly differ significantly from the evolution of the (real exact) system. A *second* numerical obstacle arises because the computer is not a formal logical system or Turing machine but a real physical machine. A physical machine is subject to physical, and especially thermodynamic, relativistic, and quantum mechanical, limitations. Of practical relevance are mainly thermodynamic limitations: Calculations, information processing, and annihilation require thermodynamic energy and produce entropy. Depending on the thermodynamic costs one is prepared to pay, different numerical limitations exist with regard to the quality of computer-generated approximation. General limitations are due to the maximum energy that is available in our cosmos for performing a calculation. Therefore, the limitation of predictions is built into the structure of the world—it is not an epistemological artefact of the way we are, nor is it just a methodological or pragmatic limitation that can be overcome. Consequently, the limitation is not "human-dependent" in any subjective sense. When dealing with unstable systems, "you [might] know the right equations but they're just not helpful," as one of the founders of fractal geometry, Mitchell Feigenbaum, argues.[33]

In other words, instabilities drive a wedge between theories (or models, equations, and laws) and predictability—with consequences for a mechanistic-deterministic worldview.[34] Despite Hume's scepticism, traditionally successful predictions of deterministic theories have been seen as the main argument in favour of a deterministic worldview. Whether an effectively reduced predictability might still provide a convincing argument for determinism remains an open question.

Three, critique of testability, confirmability, and refutability

Whereas some positions—as shown above—focus on criteria of reproducibility or predictability to characterize science, another well-established position argues that scientific propositions have to meet the criterion of testability: Both realists and empiricists—falsificationists and verificationists alike—consider empirical testability to be methodologically essential to science. Testability ensures the gaining of truth, evidence, or at least intersubjectivity; some scientists prefer to speak of objectivity in this respect.

Among others who advocated such a view, Ernst Mach (1988, 465) asserted strongly in the 1880s: "In domains where neither confirmation nor refutation does exist, we do not meet scientific standards: in fact, there is no science!"[35] For Pierre Duhem (1991, 22), the "main criterion for scientific evidence" is for a law or theory to be in "accordance with experience." On a similar note, Rudolf Carnap (1928) requires that "a statement to be meaningful" has to be based on empirical verification; "testability and meaning" turn out to be synonyms. "Verifiability"—later Carnap preferred "confirmability"—is the major criterion to demarcate science from metaphysics. Karl R. Popper (1992), from another perspective, believes in the principle of refutability. According to Popper, the growth of scientific knowledge is based on risky conjectures that (normatively: have to!) allow refutations. Central to performing refutations are the so-called *experimenta crucis* that are capable of revealing the falsehood of a proposition or theory. The existence of such crucial, decision-making experiments is indispensable to qualify a theory as a scientific one—and these experiments also show the asymmetry between the methodologically justified (because logically valid) refutation on the one hand and the methodologically unsound (because logically invalid) verification on the other hand.

The requirement of testability as a criterion to qualify science therefore is broadly shared among critical rationalists, hypothetical realists, and logical empiricists. Any type of empirical test, for both Carnap and Popper, is based on a constant relation between the mathematical-theoretical side and the empirical-experimental side. However, when dynamical or structural instabilities are present, this relation is no longer given. The behaviour of a single trajectory or a few orbits deduced from the mathematical model "cannot be compared with experiment, since any orbit is effectively non-correlated with any other and numerical round-off or experimental precision will make every orbit distinct" (Abarbanel et al. 1993, 1334). Similarly, Rueger and Sharp (1996, 103) stress: "If we test a theory in this way we will not find a precise quantitative fit, and this is to be expected if the theory is true of the system." Theory and experiment are separated into two disjunctive worlds, as Harrell and Glymour (2002) also maintain.

There are two related reasons for the issues of testability. Both the instability on the object's side and the instability on the model's side contribute to limiting testability: Because of instability, it is impossible to reproduce the object's behaviour *and* it is hard to make predictions. Either one of the problems would be hard enough to cope with, but they emerge simultaneously.

We are entering the realm of double uncertainty. For any unstable model that refers to an unstable object, Guckenheimer and Holmes (1983, 256) show that "details of the dynamics, which do not persist in perturbations, may not correspond to testable [...] properties." A classic-modern test, based on quantification and the assumption of a quantitatively constant relation, is not possible.

Four, critique of (reductive) explainability

Together with contemporaries such as Bacon or Galilei, René Descartes coined and shaped the concept of classic-modern science. The objective of science is to "trace back vague, clouded and dark propositions step-by-step to simple and evident propositions [or laws]" (Descartes 1979, 16/379).[36] Later on, in the middle of the 20th century, Hempel and Oppenheim proposed the deductive-nomological model to describe scientific explanations. This approach has become known as the covering-law model: A phenomenon is deemed explained by "subsuming it under general laws, i.e., by showing that it occurred in accordance with these laws, in virtue of the realization of certain specified conditions" (Hempel 1965, 246).[37] To put it in normative terms: Experimental phenomena have to be described by the shortcut of mathematical structure; data needs to be compressed by algorithms or laws. This claim comes close to the principle known as Ockham's razor. In this rationalist tradition, the aim of science is to achieve a minimal, simple, non-redundant description of the world. That means the challenge for physical sciences is to find laws or models as the major syntactic elements of any theory. The elimination of redundancy and the downsizing of description length are regarded as necessary conditions for explanatory success. According to Hertz (1963, xxv), "all scientists agree upon the main task: Tracing the phenomena of nature back to simple laws." This position is sometimes called explanatory reductionism or, more strongly, inter-theoretical reductionism.

Instabilities challenge reductive explanations. To some extent, this is nothing new. Early perceptions of these challenges date back to David Hume in the 18th century. Hume anticipated problems of instability in his *Inquiry Concerning Human Understanding*—although he did not explicitly use the term "instability":

> It is only on the discovery of extraordinary phenomena, such as earthquakes [...], that they [= scientists] find themselves at a loss to assign a proper cause, and to explain the manner in which the effect is produced by it. It is usual for men [= scientists], in such difficulties, to have recourse to some invisible intelligent principle as the immediate cause of that event which surprises them.
>
> (Hume 1990, 69)

Today, scientists regard earthquakes as a paradigm of unstable processes. Hume also considered earthquakes to be irregular real phenomena of nature.[38] One hundred years after Hume, Maxwell (1991, 13) pointed out that "in so far as the weather may be due to an unlimited assemblage of local instabilities it may not be amenable to a finite scheme of law at all." That is to say, "only in so far as stability subsists [...] principles of natural laws can be formulated." Only a few scientists in Maxwell's time were aware of this challenge. Nowadays, the challenge provoked by instabilities to any reductive explanation is broadly acknowledged and deeply discussed within the context of self-organization theories and nonlinear dynamics. "Nonlinear dynamical systems theory [...] studies properties of physical behavior that are inaccessible to micro reductive analytical techniques," Kellert (1994, 115) emphasizes. Where instabilities are present, we can concur with Batterman (2002) in stressing that the "devil is in the detail": Instabilities limit "approximate reasoning" and "reductive explainability."[39] In sum, unstable processes cannot be reductively condensed or compressed to a simple law.

That there exists an effective incompressibility of phenomena or data sequences generated by unstable processes is well known from information theory[40] and from chaos theory.[41] The key notion here is informational incompressibility, which is linked to essential unpredictability. According to von Neumann's ideas on self-organization, emergence, and complexity, a complex process—induced by underlying instabilities—is defined as one for which the simplest model is the realization of the process itself. The only way to determine the future of the system is to run it; there are no shortcuts and no compact laws at all. Insofar as instability underlies complexity, the simplest model of the unstable process is the process itself. Dynamical instability could also be regarded as the core of deterministic random processes and various types of non-white noise. Although these random processes might be generated by a deterministic law, it is impossible, using traditional statistical tools, to find and, further, to reconstruct the law-like structure from the data sequences or to obtain a simple law or algorithm.[42] In this vein, James Crutchfield et al. (1986, 56) stress that

> the hope that physics could be complete with an increasingly detailed understanding of fundamental physical forces and constituents is unfounded. The interaction of components on one scale can lead to complex global behavior on a larger scale that in general cannot be deduced from knowledge of the individual components.

Thus, instabilities restrict the elimination of redundancies. Unstable processes cannot be reduced to laws governing the microscopic level of atoms or molecules. Nietzsche (1930, 127) might have gone too far ahead in stating, "There are no laws of nature!"[43], but he seems to be right in stressing that there are limits to finding laws and providing explanations. An unstable world is only partly accessible and knowable. Our non-knowledge regarding the behaviour of unstable systems thus has nothing to do with a temporary insufficiency of our knowledge; it has everything to do with unstable characteristics of nature itself.

Against the Baconian position

Let us now address a further point that appears to be of central significance for the critical-reflexive account of problem-oriented interdisciplinarity. The above-discussed fourfold critique also contributes to questioning the Baconian ideal and his assumption that the *more* we acquire scientific knowledge, the *greater* will our overall capacity to control the world be.[44] In the case of instabilities, Bacon's thesis—i.e., that knowledge = control = power = (successful) technology—turns out to be a myth. Even if we were to know everything about the world we live in, instabilities do not allow us to control the things. Instabilities challenge the technical-experimental basis and also the straightforward application of scientific knowledge in the sphere of technology.

Such a finding is in sharp contrast to the Baconian standpoint. Bacon, who is widely recognized as the precursor of modern experimentalism and the ideal of technical control over the world, advocates an intervention- and action-oriented understanding of science. Bacon argues that nature should be forced by smart and well-planned experiments to unveil her nomological secrets. Later on, Kant supported Bacon in this regard and argued in favour of an action- and experimentation-oriented constructionist's (and constructivist's) point of view. To the extent that the Baconian concept of science relies on the idea about reproducibility,[45] this concept is profoundly questioned by instabilities. Even if we assume for a moment that a system obeys mathematical laws—and that lawlikeness is guaranteed (e.g., given by deterministic laws in classical mechanics)—it can elude reproducibility, predictability, and controllability. Instabilities drive a wedge between laws, on the one hand, and reproducibility and predictability, on the other hand. In sum, instabilities challenge not only experimentation but also human action in an unstable environment in general.[46]

The anti-Baconian dimension of instabilities can also be regarded from a *second*, more theoretical side, namely with regard to the limits of predictability. From the traditional, well-established perspective, predictive success is considered not only a core criterion to justify propositions, models, or theories. In addition, predictability is the very basis for any action in the world and thus for the action-theoretical foundation of engineering and technology in general.

In a nutshell, the interdisciplinary inquiry into complex systems and self-organizing phenomena, including the discovery of the prevalence and relevance of instabilities, raises questions with regard to modern conceptions of technoscientific knowledge and its control ambitions. Issues of this kind are typical for a critical-reflexive interdisciplinary approach. Since the Baconian understanding of knowledge and of nature can largely be seen as the main reason underlying the environmental grand challenges, a critique of Bacon's program can be conceived of as being central to a problem-oriented perspective.

Challenging science and philosophy of science

Let us at this point look briefly at the philosophy of science. If we accept the existence and prevalence of instabilities in nature, we are faced with not only a fourfold critique of the physical sciences themselves. In addition, traditional concepts in philosophy of science are called into question. Challenging these concepts does not have to be tantamount to completely rejecting them, but a need for rethinking them is indicated.

Instabilities, *first* of all—as we have seen—reveal limits with regard to (experimental-empirical) *reproducibility*. But experiments and the reproduction of phenomena constitute a central point of reference for many concepts in philosophy of science, such as *experimentalism, methodological constructivism,* and, to some extent, *pragmatism.* Two of the precursors of modern experimentalism are Francis Bacon and Kant, who believed that nature is experimentally constructed and conceptualized by human reason. To the extent that intervention and action-oriented concepts of science rely on strong, quantitatively oriented views about reproducibility, such concepts are called into question by instabilities.

Second, where instabilities exist, *predictability* is effectively reduced. Many philosophers who advocate an *instrumentalist* view of science uphold the predictive power of science: The success of predictions is the core argument to support their position. Concepts other than instrumentalism are also founded on arguments built around predictive success. For instance, in the tradition of Kant, *transcendental epistemologists* regard prediction as a key element of science, seeing it as a necessary condition for the possibility of acquiring knowledge—and as an unquestionable characteristic of nature: The possibility of obtaining knowledge is transcendentally guaranteed by stable, predictable, deterministic nature. According to this concept, prediction means learning from the past for the future. For *realists* and *empiricists*, predictions play an important and necessary (but secondary) role in their understanding of science. They see predictability only as a means, whereas for *instrumentalists* the objective and summit of science constitute prediction itself. Instrumentalists deem predictive success to be an indispensable criterion of justification and the basis for deciding about the truth or refutation of propositions or theories. In fact, instabilities challenge all concepts referring to predictive success.

Third, testability is called into question. Most *realists* and, from a different angle, many *empiricists* draw their arguments from the (presupposed) empirical testability of theoretically derived propositions. Hence, realists' and empiricists' concepts are based on stronger claims than reproducibility and predictability, although they draw on both. Realists and empiricists claim that the theoretical side (predictability) and the experimental side (reproducibility) must be linked: Knowledge is qualified as *scientific* knowledge insofar as theories refer to empirical propositions and can be tested. In this sense, theories approximate the empirical laws or propositions in order to provide empirical evidence for the theories. Contrary to this viewpoint, instabilities call into question the existence of such a constant relation between theory and experiment. According to Rueger and Sharp (1996, 94), "theories of nonlinear dynamics appear to generate a peculiar [...] problem for the realist [...]: in dealing with chaotic systems improved input for the theory will in general not lead to better predictions about the features relevant for confirmation."

Finally, *describability and explainability* are challenged by instabilities. *Conventionalists* and scientific *rationalists* are concerned with the descriptive and explanatory power of science. They derive their main arguments from condensed descriptions and the economy of thought. The Hempel–Oppenheim deductive-nomological scheme is the preferred type of what characterizes a satisfactory explanation. In fact, instabilities question all concepts that refer to condensed descriptions: A simple law can hardly represent an unstable time evolution.

I will not go into more detail here, but the need for further clarification is obvious. Based on a similar line of thought, Sandra Mitchell (2009) has drafted a new epistemological concept. She labels this new concept, developed in the realm of a biological approach to complex self-organizing systems, "integrative pluralism" since it faces the plurality and variety of different natural phenomena from an integrative or synthetic perspective.

The challenges to the sciences are mirrored in the challenges to the philosophy of science. Instabilities turn out to be paradigmatic for a critical-reflexive orientation of interdisciplinarity *and* of philosophy (as an academic discipline) itself.

Alternative directions of science—a third dimension of critical-reflexive interdisciplinarity

Dealing with instabilities

Although instabilities pose challenges to physical sciences, that does not render scientific inquiry and research impossible.

Throughout the history of science, scientists seem to have always had an inkling of the methodological problems provoked by instabilities. Beyond methodological considerations, the restriction of physical sciences to stable systems was also, as already discussed, supported by the underlying stability metaphysics prevalent during the historical evolution of the sciences. Stability metaphysics was undoubtedly a methodologically beneficial approach: From the perspective of the history of sciences, it is evident that the stability assumption played a crucial role in the progress and explanatory success of the physical sciences. All of this underpins that there was at least some rationale behind the implicit restriction of physical sciences to stability.

Historically, it took a long time from the first discovery of instabilities in the Newtonian era to their broader acknowledgment by the scientific community.[47] The first phase in the explicit recognition and broader consideration of instabilities dates back to the early 20th century. At that time, the challenges to sciences posed by unstable phenomena turned out to be pressing: How to deal with instabilities? Previously, the typical response to such perplexities had always been to preserve what had proven successful throughout the history of the sciences. The very last attempt to reject the challenges was to restrict sciences to the domain of stability; the methodological challenges posed by instabilities—in particular, the final effort to refute them and restrict physical sciences to stable systems—culminated in the works of Pierre Duhem and later Aleksandr Andronov.

In his seminal book *Aim and Structure of Physical Theory* (1908), Duhem pointed to works of Poincaré and Hadamard on unstable systems (Duhem 1991).[48] Even though (and probably because) Duhem was aware of instabilities, he believed physical sciences had to rely on stability. He restricted physical methodology—notably the deductive-nomological structure of explanation—explicitly to the requirement of stability. According to Duhem's methodology, deductions are theory-based predictions designed to reveal empirically accessible consequences of a physical theory (Duhem 1991, 55f). If they are to be "useful" for sciences, deductions have to take experimental uncertainties into account. Duhem's bundle concept, a key element in his hypothetico-deductive approach, comes into play here (Schmidt 2017). Specific types of non-diverging bundles reflect the experimental uncertainty of empirical facts and guarantee the application of well-established mathematical tools, such as quantitative error estimation, error theory, statistical analysis, hypothesis testing theory, and approximation theory. A diverging bundle corresponds to an unstable trajectory and for this trajectory "mathematical deduction is and always will remain useless to the physicist; [... it] will not permit us to predict what [will ...] happen" (Duhem 1991, 139).[49] Furthermore, Duhem (1991, 143) states that, when faced with the methodological issues of making deductions that will satisfy the objectives of science, the scientist is normatively "bound to impose [...] rigorous conditions" on bundles: the requirement of non-divergence or of approximability, which is the equivalent. "To be useful to the physicist, it must still be proved that the second proposition remains approximately exact when the first is only approximately true." That is, deductions are restricted to stable trajectories in order to enable predictions and guarantee the soundness of conclusions—this is a strong normative requirement, equivalent to non-divergence. Stability is a necessary condition for approximation: Deductions—if they are considered "physically or scientifically relevant"—*must* preserve the likeness of the neighbourhoods within the bundles; the second bundle must be structurally similar to the first to guarantee coping with experimental uncertainties. Scientific relevance is equated with stability. The historian Jules Vuillemin (1991, xxviii) concludes that "Duhem limits his reflections to extrapolations concerning the stability of systems." He did not deem unstable phenomena—and thus phenomena of emergence—to be located in a branch of physical sciences.[50]

In the 1930s, Andronov concurred with Duhem in restricting physical sciences to stable objects. Like Duhem, Andronov believed that science was threatened by instabilities; therefore, he was a particular proponent of (structural) stability: If stability cannot be taken for granted, it has to be imposed. As a consequence, he formulated a stability requirement which Guckenheimer and Holmes (1983, 256f) later called "stability dogma." According to Andronov, stability has to be normatively imposed in physical methodology. He asked: "Which properties *must* a dynamical system [= model] possess in order to cope with a physical problem?" The "system must be structurally stable" (Andronov et al. 1965/1969, 403f). The "stability dogma" was laid down as a strong methodological norm or requirement; it served also as a selection criterion for "physically relevant" objects.

In short, Duhem and Andronov are threshold figures indicating an age of transition. Whenever dogmatization occurs, it can be perceived as a signal of a crisis: What has until now been an unquestionable implicit belief turns out to be questionable and becomes a matter of dispute. What was formerly implicit is made explicit. Initially, the first attempt to counteract the issues raised and to refute the crisis is always to *re*-introduce dogmatically what seems to have been lost. This process of dogmatization has also been observed and described by Thomas S. Kuhn and Imre Lakatos from different angles.

Phenomenological-morphological approach

Since the 1960s at least, the Duhem–Andronov position has no longer been compelling. Yet instability is broadly acknowledged. In fact, we can observe a Gestalt switch, which not only points *negatively* to the methodological limits of the sciences but also *positively* opens avenues for alternatives in the sciences *and* in the way nature is viewed. The alternatives are linked with the advancements made in interdisciplinary structural sciences, particularly those relating to self-organization theories.

The advancements do not only foster a critical-reflexive mindset while casting doubt on the well-established mainstream of physical sciences which can be described as *classic-modern*—and which are based on the four ideals elaborated earlier: reproducibility, predictability, testability, and explainability. Self-organization theories go further; they extend and complement classic-modern sciences, for example, with regard to the objects under consideration, the methods, and the concepts. Self-organization theories, including complex systems theory, thus can be labelled *late-modern* science (Schmidt 2008a, 2015a). Late-modern science can be conceived of as a late-modern version of a phenomenological-morphological approach. "Morphology" is a term introduced by Johann Wolfgang von Goethe to signify a scientific alternative to the Newtonian methodology. On the one hand, Goethe envisioned a scientific method that is not as invasive, manipulative, and technical as the Newtonian one; on the other hand, Goethe's vision of an alternative methodology is able to address temporal processes of growth and change; it is qualitatively oriented and focuses on patterns, forms, and structures. According to Goethe, morphology can be understood as a general concept of morphogenesis—one that embraces change and growth and aims to represent nature's "Gestalt" and self-presentation. The morphological approach can be traced back, at least to some extent, to Aristotle and his Chemistry. More recent phenomenological alternatives can be found in the biology of Adolf Portmann, the medical science of Viktor von Weizsäcker, the behavioural studies conducted by Konrad Lorenz, or in the environmental studies by the von Uexkülls.

Today, the stability dogma has been replaced by an approach employing more phenomenological, qualitative, and aesthetic features relating to those properties of a theory and of an experiment which are relevant for the situation in question (e.g., specific features referred to as "complexity characteristics"). "The definition of physical relevance will clearly depend upon the specific problem. This is quite different from the original statement [...] of the stability dogma" (Guckenheimer and Holmes 1983, 259). Throughout the methodology debate on instability, it is acknowledged that weaker and context-sensitive requirements (on models/laws/theories and on experiments) are sufficient for scientific methodology and also that they are *necessary* in order to access and grasp unstable phenomena.

The weaker requirements are based on qualitative rather than on quantitative properties. For instance, Benoit Mandelbrot (1982) calls his fractal geometry a "qualitative theory of different kinds of similarities" and René Thom (1975) stresses that his catastrophe theory is a "qualitative science of Morphogenesis"—in line with some of Goethe's thinking. Morris Hirsch (1984, 11) argues that "dynamics is used more as a source of qualitative insight than making quantitative predictions." All of these works have a common background: the "qualitative theory of differential equations" which can be traced back to George David Birkhoff (1927).[51] The qualitative aspect plays a major role in the post-stability methodology. It does not confound a mathematical approach at all but contributes to further advancing a view of science conceding that science is interlaced with normativity. It opens pathways to a (context-sensitive) social epistemology in the very heart of the sciences. A key reason for this turn is that instabilities induce a strong kind of what is known as the "underdetermination thesis": Mathematical models are not underdetermined solely in a general sense by experiments and empirical measurements, as Quine and Duhem pointed out. In a stronger sense, because of the unstable nature of most objects in nature and in the social world, scientists have extensive freedom and a choice and therefore they *have to* choose; they have to decide which of the qualitative complexity characteristics of a model and of the corresponding experiment are to be seen as physically relevant, whereas the selection is more or less pragmatic and depends on the specific physical problem and on the goal of the modelling issue (i.e., on the objective, interest at stake, situation, context, resources, and the like). The choice of a particular characteristic is the starting point for any empirical test of any unstable model in the field of self-organization theories—and in late-modern sciences in general. Nancy Cartwright (1983, 1994) does not speak of characteristics but in a similar vein of capacities. Cartwright elaborates her capacity concept using the example of unstable dynamics, such as the motion of a $1000 bill from a tower down to earth. Today, an increasing amount of research is carried out on formulating, exploring, and justifying physically relevant qualitative capacities or characteristics. Not every one of these is generally suited to all problems. In certain contexts, *some* complexity characteristics are useless or meaningless; but there will be *other* characteristics enabling a classification, a detailed description, a partial explanation, or a piecewise prediction of the dynamics of a specific physical model. The variety of these contextual qualitative characteristics is striking.[52]

It deserves noting that the qualitative is not at all related to subjective impressions or to qualia-like perceptions. On the contrary, the qualitative remains mathematical but has a different meaning.[53] These qualitative complexity characteristics refer to the appearance of the phenomenon such as its shape, form, or "Gestalt" and to the geometry of the pattern *after* the process of time evolution—and not just to the bare equation. They are not grounded on *single* trajectories and *single* measurable events but on *all possible* dynamics of the *whole* dynamical system that can be depicted in the phase space (e.g., of a chaotic attractor). The global dynamics cannot be described using classical quantitative distance measures. Instead, the dynamics is characterized by means of one of the most advanced mathematical theories: differential topology and dynamical systems theory.

The idea of supplementing the quantitative focus of the mainstream methodology with a complementary, qualitative dimension is also to be observed with regard to calculability and predictability.[54] As explicated above, instabilities restrict the time horizon of single trajectory-oriented predictions. To deal with this problem, predictability is addressed explicitly and has become a major research topic in self-organization (and in complex systems) theories (Aurell et al. 1997). During the last 30 years, self-organization theorists have developed tools that accomplish qualitative calculations; the most important of these being the "shadowing lemma," which addresses issues of calculability when a quantitative approach is impossible. Using this method, we can describe the global dynamics more or less in a qualitative way. The shadowing approach provides qualitatively similar (topological) properties to characterize dynamical systems. Today, artificial intelligence methods help to extend the horizon of prediction—without any detailed knowledge about the system under consideration.

Processuality, modelling, and contextualism

The phenomenological-morphological approach is model-oriented and gives rise to different kinds of explanations. In the realm of classic-modern sciences, reductive explanations and condensed descriptions have traditionally been, and today still are, associated with universal laws and fundamental theories. As discussed earlier, instabilities put limitations on reductive explanations. That is why scientists, when accounting for unstable objects, hesitate to portray their work as discovering new theories or new laws of nature. They prefer to talk of "models" rather than of "theories" or "laws"—and regard their approach as one comprising the modelling and construction of systems and the presentation of model-based explanations. The shift in terminology is not just rhetoric; rather, it indicates a *modelling turn* in the physical sciences.[55] Traditionally, according to the standard view in the sciences, models play a minor role if any. Contrary to the standard position, René Thom (1975, 325) proposes an "Outline of a General Theory of Models." Philosophers of science, such as Hans Poser (2001, 284), go further and identify "new general types of mathematical sciences" based on "model building and simulations."[56] Stephen Kellert (1994, 85f) argues that in self-organization and chaos theory the "method of understanding" is nothing but "constructing, elaborating, and applying simple dynamical models" and "interpreting" them in different domains of application. This type of explanation referring to qualitative features of models differs from the deductive-nomological account of explanation.

A kind of contextualism seems to be the adequate position to include complexity characteristics. In the realm of unstable objects, models can be tested and justified only for a certain context of application encompassing the modelling objectives and the interests at stake, but not universally for the "whole world." It is therefore common to discuss issues such as whether or not a certain "mathematical representation or model of the system [... is] a good one" from a pragmatist perspective (Batterman 2002, 57). The respective criteria for a "good representation" are based on the requirement that the model must provide some prevalent characteristics—and not just generic ones. It is interesting that an epistemological and methodological debate on the structure of mathematical models is taking place *within* the physical sciences today—a perfect example of critical reflection on the limits and prospects of the sciences.[57]

A comparison of traditional philosophical accounts of explanation with the insights gained in self-organization theories uncovers an intriguing mismatch between the item to be explained in the former and the item to be understood in the latter. Whereas classic-modern scientists—and also philosophers of science—traditionally address scientific explanations of things such as "facts" or "events," self-organization theorists usually study behaviours, patterns, or bifurcations (cp. Kellert 1994).[58] A shift in the focus of interest, and hence in what deserves an explanation, can be observed. According to Bergé et al. (1984, 102), "the ultimate goal is to understand the origin and characteristics of all kinds of time-evolution encountered, including those which at first seem totally disorganized." The type of explanation required and appreciated in self-organization (and in complex systems) theories therefore differs from traditional, classic-modern science. Self-organization theorists favour a type of explanation "that is holistic, historical, and qualitative, eschewing deductive systems and causal mechanisms and laws" (Kellert 1994, 114). Sandra Mitchell (2009) maintains, "We are just beginning to conceptualize and to understand complex phenomena."

Critical reflection on agenda setting

Self-organization theories give rise to a shift in the focus of attention—and in knowledge interests. The simple fact that the *context of interest* precedes both the context of discovery and the context of justification is often disregarded by standard positions in philosophy of science. But the context of interest shapes any research program, such as the way in which problems are constituted or identified and objects of scientific inquiry are framed. Decisions are made explicitly or implicitly: Which objects and which phenomena are worthy of scientific research? What is of interest? What do we wish to know? What is a relevant and significant problem? In this regard, self-organization theories open avenues to reflect explicitly upon the context of interest and the purpose of knowledge.

Such reflection turns out to be necessary; decisions are indispensable. As René Thom (1975, 9) puts it, a

> choice of what is considered scientifically interesting [has to be made]. [...This] is certainly to a large extent arbitrary. [... Traditionally,] [p]hysics uses enormous machines to investigate situations that exist for less than 10^{-23} seconds [...]. But we can at least ask one question: many phenomena of common experience, in themselves trivial [...]— for example, the cracks in an old wall, the shape of a cloud, the path of a falling leaf, or the froth on a pint of beer—are very difficult to formalize, but is it not possible that a mathematical theory launched for such homely phenomena might, in the end, be more profitable for science?

Thom addresses these "phenomena of common experience" which have been "generally neglected" throughout the development of modern sciences since the 16th century (ibid., 8). The underlying reason for their neglect and disregard is clearly that "these phenomena are highly unstable, difficult to repeat, and hard to fit into a [universal] mathematical theory [...]" (ibid., 9). The objective of Thom's catastrophe theory is to find a remedy. Thom endeavours to widen the focus and to re-adjust the context of interest in order to draw attention to phenomena of common experience—phenomena that are mostly unstable, dynamical, and complex. These phenomena are at home in the familiar mesocosm of our life-world—not just located in the tiny microcosm or the huge macrocosm; we can observe them with our unaided senses. To gain access to the unstable phenomena in the mesocosm, Thom facilitates a mathematically based morphology—a "general theory of morphogenesis," including the dynamics of forms and the emergence of patterns (ibid., 101/124f).

We find cognate mesocosmic approaches in Benoit Mandelbrot's "fractal geometry" (Mandelbrot 1982), in Alfred Gierer's "physics of biological Gestalt shaping and structure formation" (Gierer 1981), and in Hans Meinhardt's "pattern dynamics of sea shells" (Meinhardt 1995). For instance, Meinhardt (1995, vii) shows that "the underlying mechanism that generates the beauty [of sea shells] is eminently dynamic. [...] A certain point on the shell represents a certain moment in its history." The "sea shell patterns are not explicable on the basis of the elementary mechanisms in a straight forward manner" (ibid.). Owing to instabilities, no quantitative (deductive-nomological) explanation is feasible, only a qualitative description based on computer simulations and graphical-visual representations. The historical time gap between the development of these tools in the 1980s and 1990s and the recognition that there was a need for them by Poincaré and others 80 years earlier can probably be attributed to the absence of adequate computational power to analyse the unstable systems. The development of computer technology has opened up new opportunities for theorizing and experimentation that enable science to deal with unstable systems. (a) *Theorizing*: The bare model equations say nothing—the phenomena are mostly hidden and cannot be foreseen. They have to be generated by numerical simulations; they are then, in many cases, represented using visualization techniques. Simulations have become an indispensable part of any theory describing unstable systems. (b) *Experimentation*: Computers not only provide new ways of dealing with theoretical problems of unstable systems—most impressive is the emergence of a *new experimental culture*; today, it is common practice to perform numerical or computer experiments (Lenhard 2007). Hence, beyond the tremendous progress on a pragmatic level, numerical experiments and computer simulations are indicative of a shift in the principles of methodology. Hans Poser (2001, 282) identifies "new criteria of scientific methodology." According to Bernd-Olaf Küppers (1993), numerical experimentations with and simulations of complex and unstable systems imply a change of "the inner constitution of physical sciences."[59]

It is now undeniable that—although, or because, the limitations of stability-oriented methodology cannot be overthrown—self-organization theories have unlocked new ways of dealing with instabilities. The classical methodological nucleus of sciences is being modified, in particular by giving more attention to qualitative than to quantitative aspects. Most striking is that phenomena, patterns, and structures within the mesocosm—and, more generally, phenomena of emergence—are arousing the interest of physical sciences.

Problem orientation—a fourth dimension of critical-reflexive interdisciplinarity

I would like to conclude this chapter by explaining why self-organization theories can serve as an excellent example of the critical-reflexive approach of interdisciplinarity presented in this book. But first, allow me to reiterate the point made earlier that the critical-reflexive approach complements the dominant instrumental approaches, which mainly address problems at the shallow end or on the surface, specifically, after they have emerged: The critical-reflexive approach facilitates a prevention of problems.

This type of interdisciplinarity concurs with Jonas and Prigogine, both of whom claim that, although the environmental problems and "grand challenges" obviously have various causes, one of them stands out because it is at the root of the others: the concept of nature being value-free and alien and subject to human disposal and control—a view that is interlaced with, or even induced by, the prevalent concept of what counts as legitimate science. Framing nature as an outer or "alien" object ("alienation of nature") which can be controlled by experimental setup and conceptualized in mathematical quantities means that any kind of value or ethical reference is just an add-on or a contingent secondary step. Such a view of nature can be seen as a core reason for the ceaseless production of new problems our societies are facing today. According to this analysis, we need a change of mindset about nature among scientists and citizens alike.

In this regard, late-modern sciences with their paradigmatic self-organization theories can contribute to changing human action towards nature. How humans see nature and the kind of ontology to which they subscribe are what determine how humans act towards nature! Admittedly, the relationship between perception and action is somewhat indirect, but it can nevertheless be regarded as being of very fundamental significance—if our aim is to change human action in the world. Jonas's line of argumentation presented in the preceding chapter might help us discern the fundamental relevance of this relationship. Since self-organization theories deliver arguments for conceptualizing nature in a different (i.e., unstable, complex, dynamic, holistic, and participatory) way, they can be regarded as problem-oriented on a deeper level: as a form of critical-reflexive interdisciplinarity that questions stereotypes and standard assumptions about nature—and about sciences.

In addition, the deficit in terms of problem orientation in the academy today is also rooted in, or at least inherently related to, what counts as legitimate science. Since late-modern sciences underscore the possibility and also the intra-scientific need for different criteria to define what counts as legitimate scientific knowledge—even in the exact sciences—they contribute to extending and widening the standards governing science and scientific expertise.[60] Late-modern sciences open pathways to a more contextual and democratic understanding of sciences.

Summary and prospect

With the advent of the paradigm of self-organization—based on the advancement of self-organization theories, dynamical systems, and complexity theories from the late 1960s on—we can observe a structural-paradigmatic shift in the physical sciences towards understanding nature as self-organizing, complex, unstable, and evolutionary. Such a view is in line with the thinking of Hans Jonas (1997, 402/18), who persistently calls for a "deep revision of how we perceive and conceptualize nature." Equipped with the concept of "self-organization" (ibid.), we can, Jonas believes, develop an alternative picture of nature that is essential to changing how we act in the natural environment.

Instabilities play a central role in this alternative view of nature, as this chapter has shown. Instabilities are a doubled-edged sword: They convey a dialectic in that they encompass a positive dimension as well as a critical and limitation-inducing dimension. On the *positive side*, instabilities constitute the source of self-organization, emergence, and the process of becoming and growth. They are the synthetic arch and unifying core of all self-organization theories, including complexity theory, nonlinear dynamics, chaos theory, dissipative structures, synergetics, and the like. As such, instabilities build a (mathematical or structural-scientific) bridge across many of the natural (and some of the social) sciences (see subsection "Synthesis—a first dimension of critical-reflexive interdisciplinarity" in this chapter). Therefore, instability-based self-organization theories can be considered a paradigmatic structural science that transcends the disciplines. More specifically, this chapter has explored ontological, epistemological, and methodological characteristics of "self-organization." Three further ontological characteristics of self-organization, in addition to (0) instabilities, are of major significance: (1) novelty, (2) processuality, and (3) internality.

The *other side* of the coin is that instabilities give rise to new criticism insofar as they challenge the methodological core of classic-modern physical sciences—including the underlying amalgam of metaphysical-methodological assumptions—such as reproducibility, predictability, testability, and reductive explainability (see subsection "Critique—a second dimension of critical-reflexive interdisciplinarity" in this chapter). Instabilities provoke intra-scientific critique and, in particular, critical reflection on how traditional sciences frame nature through the lens of stability. This critical-reflexive type of interdisciplinarity questions well-established ontological presuppositions and metaphysical convictions. As has been elaborated in this chapter, it is most interesting that the four methodological challenges to the sciences (see above) can be associated with four characteristics of "self-organization." Self-organizing phenomena show limits with regard to reproducibility/repeatability (ad 4, above), predictability (ad 5), testability (ad 6), and explainability/reducibility (ad 7). These four characteristics (ad 4–7) supplement the ontological ones (ad 0, ad 1–3); they are epistemological and methodological in nature.

Obviously, these two sides of the coin can be regarded as a dialectic relation—that is to say, the coin has a positive dimension as well as a limitation-inducing dimension. In essence, they are constitutive of what has been termed critical-reflexive interdisciplinarity: They enable synthesis as well as critique.

In addition, it has been shown that alternative sciences, sometimes named "late-modern" sciences, are emerging (see subsection "Alternative directions of sciences—a third dimension of critical-reflexive interdisciplinarity" in this chapter). Instabilities do not drive sciences into a dead end and render scientific inquiry impossible, but they engender a different concept of science and a change of view regarding what counts legitimately as science.

A further aspect also deserves to be mentioned. Conceptualizing nature as instability-based and self-organizational can be considered problem-oriented to a certain extent. Not only is such a picture of nature clearly anti-Baconian and not only does it transcend traditional dichotomies, it can also facilitate—in line with Hans Jonas's thinking—the production of ethically relevant knowledge about nature. It advances a participatory view of nature, including a different human–nature relation. An approach of this kind, which also considers the metaphysical and cultural background of the environmental crisis, is congruent with the fourth dimension of critical-reflexive interdisciplinarity (see subsection "Problem orientation—a fourth dimension of critical-reflexive interdisciplinarity" in this chapter).

In sum, instabilities turn out to be the synthetic arch across various disciplines in the field of interdisciplinary structural sciences concerned with phenomena of self-organization. Accordingly, instabilities are paradigmatic for the critical-reflexive account of interdisciplinarity—which is problem-oriented on a profound level and transgresses any account of short-range instrumentalist thinking. It thus contributes towards a kind of (self-) enlightenment in (and about) the sciences.

Most interesting is that this kind of interdisciplinarity is emerging not only beyond the disciplines but also within the disciplines themselves and is part of disciplinary advancement: It is spreading from the inside to the outside, from the part to the whole, from disciplinary science to society. Interdisciplinarity within the disciplines[61] criticizes disciplinary blindness and opens disciplines towards a broader and more-encompassing view of the things. Essentially, critical-reflexive interdisciplinarity and disciplinarity are not to be considered antagonistic.

Notes

1 In the following I will use the term "critical-reflexive interdisciplinarity" as a short version of the "critical-reflexive form of problem-oriented interdisciplinarity."
2 In consequence, it is not primarily dependent on successful method development, on the integration of knowledge, on methodology, or on stakeholder involvement.
3 In other words, considering the established conceptualization and framing of nature (including the underlying metaphysics) as well as developing and justifying alternatives are requirements for critical-reflexive interdisciplinarity. Part of this approach stands in the tradition of natural philosophy and, more recently, of philosophy of nature (cp. Schmidt 2015a).
4 This is a translation from the German version of Jonas's *Phenomenon of Life* (1997) (English version: Jonas 2001).
5 In this book, we do not assume a priority of thinking over action, as the Heideggerian quote might suggest. However, the mindset how we frame, see, and perceive nature should be taken seriously since it is central to changing our actions within (towards and with) nature.
6 My translation (J.C.S.).
7 My translation (J.C.S.).
8 This new wave of interdisciplinarity also gave rise to the notion of "late-modern sciences" (Schmidt 2008a, 2015a, 2019).

9 These limits and specific points of criticism will be discussed later on in this chapter.
10 My translation (J.C.S.).
11 My translation (J.C.S.).
12 My translation (J.C.S.).
13 My translation (J.C.S.).
14 The notion of "self organization" can be traced back to Immanuel Kant. At the end of the 18th century, Kant speaks of "self-organizing beings" in his *Critique of Judgment*. Similar ideas can be found in the works of Goethe or Schelling. Particularly Schelling embraces Kant's idea of "self-organizing beings" and envisions a new science ("speculative physics") to conceptualize a processual worldview: *natura naturans* as active, creative, acting nature.
15 The term "organization" in "self-organization" emphasizes such characteristics.
16 The same argument also holds to some extent with regard to the essential role of nonlinearity. However, nonlinearity per se does not challenge the well-established traditional methodology. Many systems are nonlinear but stable, for example, simple two-dimensional planetary systems. These, like linear systems, are without chaos and turbulence, without self-organization and symmetry breaking. Instability—and not nonlinearity—makes the difference.
17 Concepts of self-organization may differ. For instance, some of them put strong emphasis on novelty (1) but are less stringent with respect to internality (3). Besides novelty, other concepts mainly embrace irreducibility (7) and the lack of (mechanical or nomological) explanations but are reluctant to employ characteristics such as processuality (2). In sum, strong concepts of emergence require all six (above-mentioned) characteristics; weaker versions claim that only a few of them are relevant.
18 This is also explicated in Schmidt (2011a, 2015a, 2019).
19 My translation (J.C.S.). Compare an early statement by Maxwell (1991, 13).
20 My translation (J.C.S.).
21 My translation (J.C.S.).
22 The term "butterfly effect," central to the notion of "chaos," was coined by Lorenz (1963, 1989). In addition, and in more detail, over 15 competing notions and definitions of "chaos" exist (Brown and Chua 1996). One of the most relevant chaos definitions has been suggested by Devaney (1987).
23 Depending on the initial model class, altering a model can be performed by varying the model's equations or its parameters or exponents, by modifying the state space, or by altering mathematical functions.
24 Janich (1997, 62), in a similar tenor, identifies the "striking technical, prognostic and explanatory success" of physical sciences (my translation, J.C.S.). His list of characteristics might be complemented by a further aspect relating to the testability of theories in experiments and thus to falsifiability or verifiability.
25 My translation from German (J.C.S.).
26 My translation from German (J.C.S.).
27 My translation from German (J.C.S.).
28 My translation from German (J.C.S.).
29 My translation from German (J.C.S.).
30 My translation from German (J.C.S.).
31 My translation (J.C.S.). Bunge (1987) argues that nonlinearities and instabilities make it too difficult to assign determinism to unstable dynamics.
32 In the case of static instability, if the watershed is not fractal or the initial states are not located within the neighbourhood of the watershed, numerical integration is feasible. Small perturbations do not make any difference in the result or final state. In other cases, two initial states differing by imperceptible amounts may evolve into two considerably different states. Then, if any observation error of the present state is made, an acceptable prediction of the state in the distant future (to within only small tolerances) may well be impossible. Similar conclusions hold for structural instability.

33 Feigenbaum cited in Gleick (1987, 174f). The differential equation of a deterministic system is effectively worthless since reliable predictions would require exact information.

34 In particular, they drive a wedge between (deterministic) laws and (prediction-relevant, single) trajectories.

35 My translation (J.C.S.).

36 My translation (J.C.S.).

37 In the same vein, Woodward (2000, 197) claims that "explanation in the special sciences involves subsumption under general laws."

38 Hume cites the example of earthquakes to support his argumentation against anti-naturalist positions, since phenomena that appear irregular can be generated by regular laws.

39 This point is, in fact, the nucleus of the classic deductive-nomological account of explanation (see also Hooker 2004).

40 For an in-depth discussion, see, most prominently, Chaitin (1971, 1987).

41 For example, see Crutchfield et al. (1986), Devaney (1987), Peitgen et al. (1998), Schmidt (2008c, 2011a).

42 Of course, nonlinear techniques such as nonlinear data analysis, phase space reconstruction, surrogate data analysis, and other tools provide some options to find deterministic structures and separate them from white noise. These techniques are complemented by deep learning approaches in AI.

43 My translation (J.C.S.).

44 See Chapter 3 in this book.

45 These concepts are based on what is labelled "strong causation." On the contrary, instability induces only "weak causation."

46 Surprisingly, action and planning theorists have not yet perceived this challenge or tackled the issues involved.

47 The history of many nonlinear and unstable phenomena has been analysed by Darrigol (2008) and Aubin and Dalmedico (2002).

48 For an introduction to Duhem's approach to instability, see Schmidt (2017).

49 For Duhem (1991, 143), "Deduction is sovereign mistress, and everything has to be ordered by the rules she imposes" (ibid., 266). A "deduction is of no use to the physicist so long as it is limited to asserting that a given *rigorously true* proposition has for its consequence the *rigorous accuracy* of some other proposition."

50 Prigogine also points out that Duhem was pioneering in his reflection on instabilities, but that he went too far in his general critique concerning the ultimate uselessness of "unstable deductions"; Duhem assessed instabilities in a negative sense only, as being a threat to classic-modern physical methodology.

51 Some fundamental ideas are found earlier in Poincaré (1892) and in the mathematical theory of (differential) topology.

52 Some of them deserve to be mentioned: entropies, information theory parameters, Lyapunov exponents, scaling exponents, lengths of transients, fractal dimensions, renormalization theory parameters, topological characteristics, structure of stable and unstable periodic trajectories, existence of chaos or hyperchaos, parameters of basin boundaries, types of bifurcations, parameters of chaos control theory, power and Fourier spectra, phenomenological analysis of patterns. Further details can be found in textbooks on nonlinear dynamics, modelling complexity, and nonlinear data analysis (see Parker and Chua 1989).

53 Exact scientists do not require models or theories to be quantitatively stable, i.e., to show a quantitative robustness of the dynamics or structure, but to possess some complexity characteristics that are invariant under some topological transformations, that is, to be topologically equivalent or conjugate.

54 When exact scientists test a model, they need to compare the theoretically deduced model output with the data measured in the real physical-experimental system. From a methodological perspective, *nonlinear data analysis* plays a

central role in matching a model's data and the real experimental system's data (Parker and Chua 1989; Abarbanel et al. 1993). Nonlinear data analysis provides an excellent means for calculating the above-listed complexity characteristics from experimental data ("time series"). These empirically gained characteristics are confronted with theoretically deduced characteristics generated by the model. If (a) the model characteristic is proven to be prevalent in the model class and (b) the empirically measured and the model-generated data match each other within error tolerances, exact scientists will accept the model as an adequate description of the empirical phenomenon.

55 See Morgan and Morrison (1999) and Hartmann and Frigg (2006) for an introduction to the debate on the term "model."

56 My translation (J.C.S.).

57 Instability and complexity move reflection on the criteria for determining scientific evidence, truth, or objectivity into the very heart of the exact sciences.

58 This point is also elaborated in detail by Kellert (1994, 81f).

59 My translation (J.C.S.).

60 Therefore, also fields of interdisciplinary inquiry, such as technology assessment, social-ecological research, sustainability studies, or transformative research, are not to be devalued for using context dependent criteria—since this can also be observed in the physical sciences, namely in the domain of instability-based self-organization theories.

61 In this tenor, the *Oxford Handbook of Interdisciplinarity* speaks of "interdisciplinarity in the disciplines" (Frodeman et al. 2010, 77).

8 Technology and the future
Advancing prospective technology assessment

On the continuous production of pressing problems

As has been shown, we find under the umbrella of the most far-reaching, problem-oriented type of interdisciplinarity a dialectic of two main orientations: an *instrumentalist-strategic* and a *critical-reflexive* one. However, the critical-reflexive orientation of interdisciplinarity is not to be seen as an antagonistic approach but rather as one that is complementary to the instrumentalist account. To further clarify the relationship between the two orientations, we will look at a prominent field of interdisciplinary inquiry that deals with what is referred to as "real-world" problems.

Many grand challenges that societies are facing worldwide are related to environmental problems, including the highly debated sustainability and global change issues. Most of the problems have been induced by the widespread and intensive use of artefacts and technology in the global capitalist economy (Euler 1999). Since the 1960s in the US and from the late 1980s in Europe, a new interdisciplinary approach has become established and institutionalized to deal with these problems: Technology Assessment (TA). In TA projects, scholars from different disciplines such as engineering, natural and social sciences, and the humanities work together in multidisciplinary teams (Grunwald 2019). The overall objective of TA is to generate knowledge for political decision-making. This kind of policy consultancy aims to foster and facilitate the societal and political shaping of technoscientific advancement by politicians and legislation. TA counteracts the pessimistic commonplace perception of an internal momentum in the evolution of technology that is typically called "technological determinism." In TA, the advancement of technical systems is seen as being anti-deterministic; the basic purpose of TA is to identify and then to assess new technologies as early as possible—in principle in their *statu nascendi*—in order to shape their further development.

In spite of the impressive history of TA spanning more than 40 years, various concerns and points of critique have been articulated. Petra Gehring (2006), for instance, raises objections against TA and also against applied ethics in maintaining: Both (TA and applied ethics) are blind and tame because they fail to tackle the underlying self-propelling knowledge

DOI: 10.4324/9781315387109-9

dynamics of the hybrid sociotechno-economic system and its power mechanism; therefore, they are not capable of addressing the production of technoscientific problems; they are neither fundamental enough nor far-reaching enough to change the issues we face today. The objection does not only assert that TA comes too late in the day; the criticism levelled at the lateness of TA—interlaced with scepticism about the instrumentalist approaches in many variants of TA—pursues a further intention: to urge TA to address the underlying technoscientific knowledge dynamics with its inherent tendency to continuously produce new problems. That is to say, on a more fundamental level, TA is not critical enough since it fails to consider the background of the issues we face today.

Taking up this line of argumentation, I will expand the concept of TA in order to foster deeper and broader critical reflection reaching right down to the point where the problems are being created. With that goal in mind, I will sketch a critical-reflexive interdisciplinary approach in TA. Such an approach can be called *Prospective Technology Assessment* or, more precisely, *Prospective Sciences and Knowledge Assessment* (ProTA) (Liebert and Schmidt 2010). ProTA aims to facilitate self-awareness, self-reflection, and self-criticism—briefly, self-enlightenment—in the sciences and engineering, in the academy and the research system, and furthermore in science politics and society at large. An endeavour of this kind, which intends to hinder the creation of new problems, matches perfectly with the concept of critical-reflexive interdisciplinarity proposed in this book.

In general, ProTA can be regarded as being paradigmatic for the *Philosophy of Interdisciplinarity* since it is a normative-descriptive hybrid at the interface between science, society, and politics. In this sense, TA is not only interdisciplinary but also necessarily philosophic in nature.

Extending the scope

ProTA covers a broader and deeper scope than established TA concepts in that the former focuses on specific and somewhat under-exposed aspects.[1] Notably, ProTA incorporates a critical-reflexive understanding of interdisciplinarity. Its point of departure is the recognition of the inherent ambivalence of technoscientific knowledge production that is constitutive for science-based technologies in late-modern societies. Referring to this ambivalence, the orientation framework of ProTA has four components or dimensions, which will now be briefly outlined below. In addition, I will present a diagnosis of the current technoscientific situation.

ProTA considers, *first* of all, the dimension of time in the emergence of a novel kind of technology or, to be more precise, of technoscientific knowledge: it favours an *early-stage orientation*. Because we live in a science-based knowledge society, the high relevance of scientific knowledge at the beginning and throughout the entire innovation process cannot be disputed. Science and research are deemed to be the fundamental driving forces determining our societal future. While we know that innovation processes are not linear chains and have to be framed from a complex, nonlinear, and interactive perspective,[2] the relevance of initial decisions at the early stages of specific projects and programs is not to be overestimated. Consequently, ProTA has to deal mainly with science or technoscience and not just with technology and the diffusion of technical systems; it addresses the basis of our knowledge production dynamics—the root and source of the emergence of novel technical systems. ProTA therefore should be thought of more as science, technoscience, or knowledge assessment. The early-stage approach of ProTA goes right to the epicentre of academic knowledge production and is characteristic of a critical-reflexive account. Seen from that angle, the traditional idea (and rhetoric) of technoscientific knowledge being value-free becomes nothing but a myth.

Second, because the consequences and impacts of technologies are hard to anticipate in concrete terms at the early stages, ProTA addresses purposes, intentions, potentials, and visions. Of central interest here are the consideration of technoscientific (realistic) potentials and their demarcation from (unrealistic) visions, promises, and hypes. Because pure basic research and purposefully applied research are today highly intermingled,[3] purposes migrate far into science and laboratory practice. Science in its inner constitution is purpose-driven and strongly value-laden. Accessing and assessing purposes, including the options for realizing the potentials, are key elements of ProTA. We know—or at least could know—much at the very beginning of science and research processes: during the phases of agenda setting and the development of research corridors. In principle, negative side effects and risks can be identified very early on. An early anticipation of unintended consequences is feasible—without obtaining strongly prognostic-predictive knowledge. There is an obvious reason for this. Many present-day and future technologies are based on predecessor technologies or on a synergetic combination of already-established technologies. The knowledge about these is already present, for instance, in the field of future nuclear technology. Thus, ProTA is concerned with the state-of-the-art in the sciences or technosciences. On this basis, it discloses the various kinds of non-knowledge on different levels (uncertainties, ignorance, and risks) and it considers possible non-knowables. It is guided by the precautionary principle that has been adopted in order to deal with uncertain but nonetheless relevant knowledge (Manson 2002). In sum, ProTA facilitates public and intra-scientific discourse on the intentions, potentials, and visions of a novel or expanded technoscientific field or both. It revisits, reflects on, and potentially revises the purposes and potentials at all stages of the technoscientific innovation process.

Third, ProTA is *shaping-oriented*. As such, it initiates reflexive search procedures to find and assess alternative paths and trajectories of technoscientific advancement. To accomplish this approach, it does not stand for an uninvolved observer's perspective, as has been predominant in classic concepts of expertise and policy advice—a typically instrumentalist viewpoint. To the contrary, in creating knowledge, ProTA brings together those involved in the overall process of developing, designing, shaping, and moderating technoscientific knowledge. In other words, the addressees as well as the participants in ProTA are also those who are actively involved in, or contribute to, the shaping process: scientists, engineers, politicians, university leaders, academy administrators, and program managers. Overall, the shaping orientation is based on an analysis and critique of the current state of the joint science, technology, and innovation system. The analysis reveals that the scientification and technization of society and the life-world in conjunction with the complementary socialization of science and technology are central features of late-modern societies. The optimistic faith in science and technology prevalent in the 1960s and 1970s, which held that these processes can be intentionally planned, steered, and effectively controlled, is gone. Science, technology, and society are intertwined in a great variety of ways. New concepts and approaches, such as governance, engender hope that—in lieu of steering the production of technoscientific knowledge— shaping the procedures could improve scientific–societal co-activity. They also call into question the externalist perspective and the assumptions of classic, control-oriented action theory, which are most prominent in the instrumentalist notions of prediction and control. Based on such an analysis of the status quo, the point of departure of ProTA is the observation that the boundary between the intra-scientific and extra-scientific realms—in other words, between the academic system and society—is becoming blurred. In this regard, ProTA can be seen as a kind of engaged and participatory research to shape technoscientific knowledge production.

Fourth, ProTA focuses on the technoscientific core of an emerging or novel type of technology. It is an assessment based on a detailed examination of the technoscientific knowledge envisioned or already available in the field under consideration—which includes the feasible or already realized technical systems, the experimental setups, and the technical artefacts and infrastructures. Scientific and engineering knowledge is very much required in order to enable an analysis of this kind. Scientists and engineers have to become involved when a ProTA is carried out, for example, in the political arena concerning novel research and development programs. Although ProTA focuses on the technoscientific core, it does not do so in a narrow sense. On the contrary, it is strongly embedded in the societal sphere. For instance, societal values and visions play a central role in the pursuit of technoscientific knowledge and thus in the development of the technoscientific core. ProTA uncovers the values driving visionary technical design processes and, if necessary, criticizes them. Consequently, consideration of the deeper knowledge produced at the technoscientific core is indispensable. The status of current research and the intra-scientific dynamics may also be scrutinized to clarify the extent to which there is potential for the visions to be fulfilled. Addressing and assessing the technoscientific core—as central elements of future-relevant discussions regarding technoscientific potentials, intended impacts, expected results, and non-intended consequences— are really what make ProTA a truly interdisciplinary enterprise.

Thus, ProTA provides a fourfold orientation for a critical-reflexive approach. It focuses on the early stages of a development (first point above), on purposes, intentions, visions, and potentials (second) related to technoscientific knowledge in order to (fourth) enable alternatives in the design and shaping (third) of the novel technoscientific knowledge under consideration.

Normative anchor

In addition to the fourfold orientation framework, ProTA—as a critical-reflexive kind of interdisciplinary practice—explicitly reflects on its normative fundament. The fundament can be delineated by referring to two criteria based on philosophical considerations (Liebert and Schmidt 2010).

The first criterion was developed by Hans Jonas (1984) in his *Imperative of Responsibility* and is related to the heuristics of fear and to a prudent manner regarding action.[4] We can say that Jonas formulated a *preservation principle* aimed at achieving a conservative preservation of our life-world and genuine human life. The second, somewhat complementary criterion is related to Neo-Marxist thinker Ernst Bloch's (1995) utopian *Principle of Hope*, which addresses the open horizon of the desired future. In his anti-dualist materialist dialectics, Bloch sees man and the world as unfolding in a non-teleological way. He envisions what he calls an *alliance technology* that serves mankind *and* is concurrently in resonance and harmony with nature. Let me now elaborate somewhat on the twofold normative framework.

Jonas uses the term non-reciprocal to describe responsibility. The crucial aspect here is not the relationship between equals but the asymmetric relation: the responsibility *for* somebody or something, *for* others. It is a responsibility for other humans, for future generations, for animals, for embryos, or for nature. The concept of non-reciprocity encompasses the responsibility *of* scientists *for* other human beings (i.e., those who are affected by research outcomes) and *for* the natural environment. The "so-being" of the world calls us to preserve it. We therefore can ask in concordance with Jonas: Do our approach to nature and our ways of producing knowledge contain ethical core elements to secure the permanence of "genuine human life"? That is not the case: According to Jonas, we have to shift ethical reflection to an earlier phase, where scientific knowledge is generated. Ethics and (scientific) knowledge should not be deemed separable from each other. Societal responsibility has to be part of the whole research process from the very beginning.

Now let us look at a complementary approach offered by Bloch's *unfolding principle*. According to Bloch, what needs to be done is to positively define the envisioned directions of technoscientific progress or, in other words, to *unfold* and to constitute the societal future we desire. The current *so-being* with its deficiencies and problems is the point from which the unfolding principle takes its departure. With its strict, forward-looking perspective, the unfolding principle is aimed at enabling decisions on what is *desirable* and *possible* beyond what already exists: How would we like to live in the future while acknowledging the plurality of different ways of life? To answer this central question, we need to consider both the values behind our wishes for the future and the technoscientific options for moving in the desired direction. A prerequisite, as we have seen earlier, is to address and assess the technoscientific core. Such an assessment relates to the desired future; it is a value-sensitive process for which explicit reflection on normative backgrounds is indispensable. Some dimension of this kind of approach towards constituting the desired future and the unfolding potential can be identified in the field of discourse ethics (Habermas 1993). Discourse ethical procedures claim to form a fundament of deliberative considerations of this nature within future-oriented goal-setting processes: In a power-free discourse, interlaced with a mutual recognition of interests and sense of values, we can debate and examine what is universally acceptable and what is not. The discourse must not necessarily lead to a consensus. Well-founded and transparent dissenting opinions could also prove helpful and serve as a basis for individual, institutional, and political decisions, which should always be structured in a reversible manner.

In addition to elements of deontological and virtue ethics (Jonas) and of discourse ethics (akin to the thinking of Bloch), utilitarian concepts are indispensable to the concept of ProTA. The consideration of consequences is inherent to utilitarian approaches. It is naturally of relevance for any kind of TA which enquires into intended and unintended effects and outcomes. The significant utilitarian arguments are those underlining that good intention and convincing justifications do not suffice. At the same time, putting emphasis on the outcomes implies—conversely—that the motives, intentions, visions, and interests (in relation to scientific action) are somewhat under-exposed. A major drawback of the utilitarian approach of choosing between conflicting benefits is that it can prevent us from considering and posing fundamental questions. Moreover, utilitarian perspectives often promote the seizing of alleged chances in the absence of proven risks and thus there is an emerging tendency to annul the balance between the principles of preservation and unfolding that is central for ProTA.

To summarize, ProTA incorporates elements of the most common ethical concepts. It fills deontological concepts with material-normative or virtue-ethical content—some would call it "metaphysical"—which is related to the *preservation principle* proposed by Hans Jonas. Discourse ethics is central insofar as it enables a deliberative goal- and vision-setting process that can be linked to Bloch's *unfolding principle*. In addition, utilitarian-consequentialist thinking is—at least to some extent—indispensable in order to appropriately include outcomes, consequences, decisions, and actions pertaining to concrete research programs. In any event, ProTA can hardly avoid being underpinned by concepts of ethics.

Synthetic biology – a case study

To elaborate the discussion further, I will present an example of ProTA in the field of synthetic biology. This field is reputedly a key technoscience of the future.

In 2010, the research entrepreneur Craig Venter announced the forth-coming advent of an epochal break and envisioned a fundamental shift in our technical capabilities. Synthetic organisms

> are going to potentially create a new industrial revolution if we can really get cells to do the production we want; [...] they could help wean us off of oil, and reverse some of the damage to the environment like capturing back carbon dioxide.
>
> (Venter 2010)

Venter's visionary claim was evidently induced by the success of his team in the *Creation of a Bacterial Cell Controlled by a Chemically Synthesized Genome*—as his article in *Science* magazine was titled (Gibson et al. 2010).

In fact, the hype that Venter generated has actually set off another huge wave. He has been accused of "playing God" or, at least, of advocating a dangerous type of "hubris." Although such concerns and objections to Venter's optimism are key elements in the formation of public opinion and political deliberation, both extreme positions—Venter's and that of his critics—often lead to a deadlock. Maintaining them would mean missing opportunities to engage in shaping this new technoscientific wave. In this respect, the concept of ProTA enables, from a critical-reflexive perspective, an earliness approach providing relevant background information for ana-lysing and assessing the new technoscientific wave (Liebert and Schmidt 2010). ProTA, as we have seen, can be regarded as an endeavour that extends and expands established TA concepts by focusing on specific and somewhat underexposed elements in the existing concepts. It is based on a fourfold orientation framework, as described above: an earliness orien-tation, an intention orientation, a shaping orientation, and an orientation towards the technoscientific core.[5]

In the following, I will examine in more detail how the fourth orientation of ProTA, which is certainly intertwined with the other three, applies to the field of synthetic biology. The thesis presented here is that the major essence of the technoscientific core of "synthetic biology" is the idea(l) of harness-ing self-organization—including the ability to set off complex dynamical phenomena—for technical purposes.

Synthetic biology is still in its infancy. The societally relevant ethical issue demanding consideration at this early stage in the technology's development is that, should a technology based on self-organization ever be attained and implemented, we would enter a new technological era in which technical systems possessed high levels of autonomy and agency properties. The risks would be hard to assess. The systems would "take on a life of their own such that we no longer appear to perceive, comprehend, or control them" (Nordmann 2008b, 176). The new type of technology could be called "late-modern," indicating that it is ontologically different from, and an extension of, the recent modern kind of technology.

Scrutinizing the visions

What does the technoscientific core of synthetic biology consist of, and what is the common denominator of synthetic biology? The exact meaning of the umbrella term "synthetic biology" is, in fact, not at all clear. New labels and trendy watchwords generally play a key role in the emergence of new technoscientific waves. Synthetic biology is certainly no exception in that it is an extremely popular buzzword widely encountered in debates on research politics, as was the notion of nanotechnology more than a decade ago.[6]

All TA scholars and ethicists are aware of the fact that labels are strongly normative. Labels are not innocent or harmless. They carry content and form the backbones of visions. They are roadmaps towards the future and can quickly turn into reality; they shape the technoscientific field and determine our thinking, perception, and judgment. Labels help to foster hopes and hypes as well as concerns and fears; their implicit power to create or close new research trajectories and development roadmaps can hardly be overestimated. Labels are part of what could be described as "term politics," regulating and shaping the specific field with a "gatekeeper function" that decides who is *in* and who is *out*; whose research field can be deemed "synthetic biology" and whose is merely a subfield of traditional biotechnology. Labels are relevant with respect to funding, publication opportunities, reputation, and career. Thus, they determine and sway our future in one way or another. To what does the umbrella term "synthetic biology" refer? Is there a unifying arc? What visions do synthetic biologists have, and how likely will their visions be achieved? Three popular visions or definitions[7] of synthetic biology stand out.[8]

First, the *engineering vision* frames synthetic biology as being radically new since it is said to bring an engineering approach to the scientific discipline of biology. This vision is governed by the ideal of making new genomes or transforming existing genomes by the insertion of new genes/gene sequences or by the elimination of existing genes. An engineering understanding is advocated by a High-Level Expert Group of the European Commission: "Synthetic biology is the engineering of biology: the synthesis of complex, biologically based (or inspired) systems [...]. This engineering perspective may be applied at all levels of the hierarchy of biological structures [...]. In essence, synthetic biology will enable the design of 'biological systems' in a rational and systematic way" (European Commission 2005, 5). This comes close to the definition given by Pühler et al. (2011), who see synthetic biology as "the birth of a new engineering science." Similarly, others view synthetic biology as "an assembly of different approaches unified by a similar goal, namely the construction of new forms of life" (Deplazes and Huppenbauer 2009, 58). The engineering definition is generally based on the assumption that before synthetic biology arose, there was a clear dividing line between biology as an academic discipline, on the one hand, and engineering/technical sciences, on the other. Biology is regarded as a pure science aiming at fundamental descriptions and explanations. In contrast, engineering sciences appear to be interested primarily in intervention, construction, and creation. Seen in that light, biology and engineering sciences have traditionally been perceived to be—in terms of their goals—like fire and ice. The proponents of the engineering definition believe that the well-established divide between the two disciplines is becoming blurred. Today, engineering goals are being transferred to the new subdiscipline of biology. According to the advocates of this definition, these goals have never been characteristics of other subdisciplines of biology (*divergence from traditional biology*). The essential claim is that we are experiencing an epochal break or a qualitative shift in the aims and approaches of biology as well as in how the field is understood: In this light, biology is aimed not at theory but at technology. Synthetic biology appears to epitomize the ideal of the technoscientification, technicization, or engineering of biology.

Second, the *artificiality vision* in regard to synthetic biology is related to the former definition but is concerned more with objects than with goals. According to the European Union (EU) project TESSY ("Towards a European Strategy for Synthetic Biology"), synthetic biology deals with "bio-systems [...] that do not exist as such in nature" (TESSY 2008). In an equivalent sense, others have stated that synthetic biology encompasses the synthesis and construction of "systems, which display functions that do not exist in nature" (European Commission 2005, 5). The German Science Foundation, together with the Academy of Technical Sciences and the National Academy of Sciences Leopoldina, similarly identifies the emergence of "new properties that have never been observed in natural organisms before" (DFG et al. 2009, 7). It defines synthetic biology by the non-naturalness or unnaturalness of the constructed and created bio-objects. *Divergence from nature* appears to be the *differentia specifica* of synthetic biology, and nature is seen as the central anchor and negative foil for this definition. Whereas bio-systems were traditionally natural (i.e., they occurred exclusively *within* and were created *by* nature alone), the claim here is that, from now on, bio-systems can also be artificial (i.e., created intentionally by humans). That is certainly a strong presupposition, which is also linked to the idea of a dichotomy between nature and technical objects. The dichotomy can be traced back to the Greek philosopher Aristotle, who drew a demarcation line between *physis* (nature) and *techné* (arts and technical systems). In spite of Francis Bacon's endeavours at the very beginning of the modern epoch to eliminate the dichotomy and naturalize technology, the nature–technology divide broadly persists in the above definition. In a certain sense, the artificiality definition of synthetic biology presupposes the ongoing plausibility of the Aristotelian concept of nature, neglects the Baconian one, and argues for an epochal break in understanding bio-objects and bio-nature: These are not given, they are fabricated.

Third, the *extreme gene technology/biotechnology vision* leads either to synthetic biology being seen in a more relaxed light or, on the contrary, to its being condemned as a continuation of trends already perceived as terrible and dangerous in the past. According to the proponents of this definition, we are experiencing just a slight shift and mainly a continuation, not an epochal break; nothing is really new under the sun. Synthetic biology merely extends and complements biotechnology. Drew Endy (2005, 449), a key advocate of synthetic biology, perceives only an "expansion of biotechnology." Similarly, but from a more critical angle, the Action Group on Erosion, Technology and Concentration (2007) defines synthetic biology as an "extreme gene technology," mainly because it is based on gene synthesis and cell techniques such as nucleotide synthesis, polymerase chain reaction, or recombined cloning. The basic methods, techniques, and procedures have been well established since the late 1970s. Although there have been tremendous advances from a quantitative standpoint, it is hard to discern any qualitative progress in the core methods. The extreme biotechnology definition rarely deals with goals or objects, but with methods and techniques. Its proponents claim (1) that, implicitly, methods constitute the core of synthetic biology, (2) that there has been no breakthrough in the synthetic/biotechnological methods, and moreover (3) that a quantitative advancement cannot induce a qualitative one. Briefly, this position perceives a *continuation in methods*—in contrast to a divergence from biology or nature as perceived in the former two definitions.

We are faced with a plurality of three different conceptions of what "synthetic biology" means or, speaking in normative terms, what it should mean. The three visions or definitions—the engineering, the artificiality, and the extreme biotechnology vision—tell three different stories. Each one exhibits some degree of plausibility and conclusiveness. In spite of their apparent differences, all are concerned (first) with disciplinary biology or biological nature and (second) with a rational design ideal in conjunction with a specific understanding of technology, technical systems, and engineering action. However, that is not the whole story.

First, the focus on biology as a standalone discipline, including a discipline-oriented framing, prevents an exhaustive characterization of the new technoscientific wave: Synthetic biology is at its nucleus far more *interdisciplinary* than disciplinary. This point needs to be taken into account when seeking an adequate definition: Biologists, computer scientists, physicists, chemists, material scientists, medical researchers, and people from different engineering sciences are engaged in synthetic biology. Because various disciplinary approaches, methods, and concepts coexist in synthetic biology, the term seems to serve as a label for a new interdisciplinary field. Accordingly, a biology bias would surely be overly simplistic and entirely inadequate; framing synthetic biology as merely a new subdiscipline of biology would represent a far too narrow approach. It is not sufficient to provide a clear understanding of synthetic biology. Thus, we need to ask whether we are dealing with a much more fundamental technoscientific wave than simply a change in one particular discipline or academic branch.

Second, in line with what is referred to as bionano or nanobio research, the three definitions look at synthetic biology from the angle of technology and engineering. This manner of approach appears plausible in some respects: Synthetic biology extends and complements advancements in nanotechnology and hence spurs a position that can be called "technological reductionism" (Schmidt 2004, 35f). Technological reductionists aim to eliminate the patchwork of engineering sciences by developing a fundamental technology or a "root, core, or enabling technology" (ibid., 42). The slogan fostered by technological reductionists is: shaping, designing, constructing, and creating the world "atom-by-atom." Eric Drexler is a prominent advocate of technological reductionism. He argues that there are

> two styles of technology. The ancient style of technology that led from flint chips to silicon chips handles atoms and molecules in bulk; call it bulk technology. The new technology will handle individual atoms and molecules with precision; call it molecular technology.
>
> (Drexler 1990, 4)

Interestingly, recent technological reductionism ("molecular technology"), Drexler upholds, complements and perfects the traditional ("bulk") technology. The three definitions of synthetic biology described above concur strongly with technological reductionism; it certainly seems plausible to put synthetic biology in the context of this new type of technology-oriented reductionism. But whether that is all that can, or should, be said to characterize synthetic biology remains to be clarified. It is absolutely clear that synthetic biology differs from nanotechnology, which can be viewed as a paradigm of a technological reductionist approach (Schmidt 2004). Many synthetic biologists claim to pursue an approach that is complementary to nanotechnology and has been called a "systems approach" or, in a more visionary sense, "holistic." Given the widespread reference to "system," along with the alleged successful application of "systems thinking," synthetic biology seems to involve a convergence, or dialectical relationship, of seemingly contradictory concepts: (systems) holism and (technological) reductionism with its strong control ambitions and emphasis on rational engineering. This inherent dialectic is obviously central to an appropriate understanding of synthetic biology. The three definitions presented so far do not encompass this point.

In light of that omission, our characterization of synthetic biology (and its technoscientific core) has to go beyond the three narrow definitions given above. Although it is neither erroneous nor misguided to see synthetic biology (i) as a subdiscipline of biology and (ii) as a technologically reductionist position, this conception is one-sided, biased, and limited in depth and scope. It needs to be supplemented with an approach that also takes fundamental tendencies in science and technoscience in general into consideration and focuses in more detail on the technoscientific core of the emerging technoscientific wave.

Deepening the analysis

Synthetic biology aims to harness self-organization for technical purposes

To arrive at a more fitting and more comprehensive characterization of synthetic biology, we should not restrict ourselves to goals (as in definition 1), to objects ("ontology," as in definition 2), or to methods ("methodology," as in definition 3) but also consider the underlying principles and concepts within the technoscientific field, namely the technoscientific core. This requirement is central to the approach of ProTA. Thus, we need to include a fourth definition—*the systems or self-organization definition*—that is prevalent in synthetic biology research programs.

Synthetic biology makes use of the self-organization power of nature for technological purposes: "Harnessing nature's toolbox" in order to "design biological systems," as David A. Drubin, Jeffrey Way, and Pamela Silver (2007) state. Even back in 2002, before synthetic biology had been broadly discussed (although its main ideas were already on the table), Mihail Roco and William Bainbridge (2002, 258) *anticipated new frontiers in research and development by* "learning from nature." They perceived the possibility of advancing technology by "exploiting the principles of automatic self-organization that are seen in nature." According to Alain Pottage and Brad Sherman (2007, 545), the basic idea of synthetic biology is to "turn organisms into manufactures" and to make them "self-productive." The paradigm of self-organization and self-productivity is implicitly or explicitly articulated in many papers on synthetic biology. Pier Luigi Luisi and Pasquale Stano (2011) also advocate an understanding of synthetic biology based on self-organization:

> Synthetic cells represent one of the most ambitious goals in synthetic biology. They are relevant for investigating the self-organizing abilities and emergent properties of chemical systems—for example, in origin-of-life studies and for the realization of chemical autopoietic systems that continuously self-replicate—and can also have biotechnological applications.

Jean-Pierre Dupuy (2004, 12f) discerns that "[t]he paradigm of complex, self-organizing systems is stepping ahead at an accelerated pace, both in science and in technology." Jordan Pollack puts self-organization at the very centre of his vision of designing advanced biomaterial. Pollack's goal is to "break [...] the limits on design complexity," as his article is entitled. "We think that in order to design products 'of biological complexity' that could make use of the fantastic fabrication abilities [...], we must first liberate design by discovering and exploiting the principles of automatic self-organization that are seen in nature" (Pollack 2002; in Roco and Bainbridge 2002, 161).[9]

In fact, the systems approach of putting the self-organization power of bioengineered entities at the very centre of the new technoscientific wave has enjoyed an impressive evolution over the last three decades. It goes back to one of the most popular and highly controversial publications by K. Eric Drexler in the early 1980s. Drexler talks about "self-assembly," "engines of creation," and "molecular assemblers." "Order can emerge from chaos without anyone's giving orders [... and] enable[s] protein molecules to self-assemble into machines" (Drexler 1990, 22f.). "Assemblers will be able to make anything from common materials without labor, replacing smoking factories with systems as clean as forests." Drexler goes further and claims that emergent technologies "can help mind emerge in machine." Richard Jones (2004) takes up Drexler's ideas and perceives a trend towards self-organizing soft machines that will change our understanding of both nature and technology. From a different angle but in a similar vein, the 2009 report "Making Perfect Life" of the European Technology Assessment Group (2009, 4) refers to advances in synthetic biology: "Synthetic biology [...] present[s] visions of the future [...]."

> Technologies are becoming more 'biological' in the sense that they are acquiring properties we used to associate with living organisms. Sophisticated 'smart' technological systems in the future are expected to have characteristics such as being self-organizing, self-optimizing, self-assembling, self-healing, and cognitive.[10]
>
> (ibid.)

Alfred Nordmann (2008b, 175) sees a new understanding of technology emerging in the field "where engineering seeks to exploit surprising properties that arise from natural processes of self-organization." A "shift from" what Nordmann describes as "nature technologized" to "technology naturalized" can be observed, which "is usually hailed as a new, more friendly as well as efficient, less alienated design paradigm" (ibid., 175).

Synthetic biology—it is interesting to observe—does not stand alone: Self-organization also plays a constitutive role in other kinds of emerging technologies such as

1. Artificial intelligence, machine learning, robotics, autonomous software agents, and bots;
2. Nano- and nanobio-technologies;
3. Cognitive and neuro-technologies.

Moreover, self-organization in technical systems serves as a leitmotif in science policy: "Unifying science and engineering" seems possible by "using the concept of self-organized systems" (Roco and Bainbridge 2002,

10/84). Self-organization appears to be the conceptual kernel of the ideal of the *convergence of technologies* and also seems central to any kind of *enabling technology* (ibid.; Schmidt 2004). The above list of examples shows that synthetic biology is not unique; it can be perceived as being only a prominent example or as the spearhead of a universal trend in technology.

Synthetic biology as late-modern technology

If we take the visionary promises as serious claims, they announce the emergence of a new type of technology. We do not know whether the promises can be fully kept. However, if this were to be the case, we would encounter a different kind of technology, including novel risk issues having ethical relevance: a *late-modern technology*.

Late-modern technology has nothing to do with our established perception of traditional technical systems. It shows nature-like characteristics; it does not present the appearance of being technology; it seems to be "un-technical" or "non-artificial"; the signs and signals, the tracks and traces of technology are no longer visible (Hubig 2006). Technical connotations have been peeled off; well-established demarcation lines are blurred. Late-modern technology seems to possess an intrinsic momentum of rest and movement within itself—not an extrinsic one. Such characteristics come close to the Aristotelian and common life-world understanding of nature: Technology *is* alive, or *appears* to be alive, as nature always has been. The internal dynamics (i.e., activity, change, and growth) of self-organization technology make it hard to draw a demarcating line between the artefactual and the natural in a phenomenological sense: Nature and technology seem indistinguishable. Even where it is still possible to differentiate between the artificial and the natural (e.g., in robotics), we are confronted with an ever-growing number of artefacts displaying certain forms of behaviour that traditionally have been associated with living systems. The words used by Schelling and Aristotle to characterize nature also seem to apply to technology: A late-modern technical system is "not to be regarded as primitive" because it appears to act by itself: (a) it creates and produces, (b) it selects means to ends, and (c) it makes decisions and acts according to its environmental requirements. Technology evidently presents as an acting subject: "Autonomy"—a term central to our thought tradition—is ascribed to these systems.

What is behind this trend towards a *phenomenological convergence* of nature and technology or, in other terms, towards the *phenomenological naturalization of technology*—apart from "technological reductionism"? To answer this question, we need to examine the claims made by the advocates of synthetic biology. Far more relevant and foundational, it seems, is the aspect we could call *nomological convergence*, which engenders a fundamental trend towards the *nomological naturalization of technology*. Mathematical structures describing self-organization in technical systems are said to converge with those in nature. Although the objects might differ, their behaviour and dynamics show a similarity. According to M.E. Csete and J.C. Doyle (2002, 1664), "advanced technologies and biology are far more alike in systems-level organization than is widely appreciated." The guiding idea(l) of nomological convergence dates back to the cyberneticist and structural scientist Norbert Wiener. He defined structure-based convergence with regard to specific "structures that can be applied to and found in machines and, analogously, living systems" (Wiener 1968, 8).[11] The physicist and philosopher Carl Friedrich von Weizsäcker pointed out some 50 years ago:

> Structural sciences encompass systems analysis, information theory, cybernetics, and game theory. These concepts consider structural features of different objects regardless of their material realm or disciplinary origin. Time-dependent processes form a common umbrella that can be described by an adequate mathematical approach and by using the powerful tools of computer technology.[12]
>
> (Weizsäcker 1974, 22f)

Today, we can add self-organization theories which encompass nonlinear dynamics, complexity theory, chaos theory, catastrophe theory, synergetics, fractal geometry, dissipative structures, autopoiesis theory, and others. Following the first wave of structural and systems sciences such as information theory, game theory, and cybernetics (Bertalanffy, Wiener, Shannon, and von Neumann) in the 1930s and 1940s, we are now experiencing a second wave (Maturana, Varela, Prigogine, Haken, Foerster, Ruelle, and Thom) that began in the late 1960s. Self-organization, macroscopic pattern formation, emergent behaviour, self-structuring, growth processes, the relevance of boundary conditions, and the Second Law of Thermodynamics (entropy law) with its irreversible arrow of time are regarded as conceptual approaches to disciplinarily different types of objects, based on evolutionary thinking in complex systems. Assisted by the spread of computer technology, concepts of self-organization had a tremendous impact on scientific development in the second half of the 20th century.

Tracing the technoscientific core

The thesis proposed in this chapter is that synthetic biology harnesses, or aims to harness, self-organization capability for technical purposes. However, the term "self-organization" is not very precisely defined. Since Kant's and Schelling's coining of "self-organizing beings," the concept of self-organization has been in flux, although the term seems to have retained its essential meaning, which is the immanent creation and construction of novelty:

- the emergence of novel systemic properties—new entities, patterns, structures, functionalities, and capacities.

Notwithstanding the philosophical debate on the notion and characteristics of novelty, the following are widely accepted criteria to specify "self-organization":[13]

- dynamics, processes, time-dependency, and historicity;
- internality or "autonomy" (the notion of "self" in "self-organization");
- irreducibility of the description;
- unpredictability of the self-organized or emergent phenomena.

In consequence, self-organization processes are generally non-separable from their environment; they are hard to control by an external actor. "The engineers of the future will be the ones who know that they are successful when they are surprised by their own creations" (Dupuy and Grinbaum 2006, 289). In brief, the notion of self-organization is, from an engineering perspective, linked to characteristics such as "productivity/creativity," "processuality," and "autonomy." These terms are frequently used by synthetic biologists.

I have stated that harnessing self-organizing power for technological purposes is at the core of synthetic biology. But what is at the core or root of self-organization? Basically, the answer I propose is that *instabilities* turn out to be essential for self-organization; they are constitutive to all systems or structural theories (cp. Schmidt 2011).[14] According to Gregory Nicolis and Ilya Prigogine (1977, 3f), "instabilities are necessary conditions for self-organization." As seen in the previous chapter, instabilities are generally situations in which a system is on a razor's edge: criticalities, flip or turning points, thresholds, and watersheds. They generate sensitive dependencies, bifurcations, and phase transitions. The classic-modern *strong* type of causation does not govern these processes; rather, it is the *weak* causation that enables feedback procedures and amplification processes. Instabilities can induce random-like behaviour, deterministic chance, and law-based noise, which are inherently linked to uncertainty. The most prominent example to illustrate instability is the "butterfly effect." The beating of a butterfly's wings in South America can have a tremendous impact on the weather in the US and cause a thunderstorm (Lorenz 1963; Lorenz 1989; Schmidt 2011a).

Unstable systems show certain limitations with regard to predictability, reproducibility, testability, and reductive explainability. An isolation or separation of the systems from their environment is impossible because they are continuously interacting with it. In general, instability should not be equated with the collapse of a system. Insofar as present-day engineers intend to make use of self-organization power, they have to provoke and stimulate instabilities: Self-organization requires that a system's dynamics pass through unstable situations. To put it metaphorically, late-modern technology can be considered the technoscientific attempt to initiate and stimulate a *dance on the razor's edge*. This specific, highly sensitive technological core is the basic object of interest for accomplishing an early assessment;

and it is at the very centre of the concerns raised by Hans Jonas, who was a precursor in anticipating an instability-based complex technology.

Assessing the technoscientific core

The instability-based type of technology is somewhat ambivalent because it obviously carries an internal conflict or considerable dialectic that cannot be overcome by minor modifications of the technical system itself.

On the one hand, instabilities constitute the core of self-organization and hence of technologically relevant self-productivity. On the other hand, instabilities are intrinsically linked to obstacles and limitations, not only with regard to the construction and design accomplished by the technical systems but also with regard to the possibility of subsequently controlling and monitoring the systems. When instabilities are present, tiny details are of major relevance; minor changes in some circumstances can cause tremendous, unforeseeable effects; unstable systems lack predictability. Owing to empirical-practical and fundamental-principle uncertainties, the tiny details are hard to control. Paradoxically, although they are constructed by humans, the systems remain fundamentally inaccessible and elude comprehension and control (Nordmann 2008b; Köchy 2011).

On account of these limitations, technology and instability were traditionally like fire and ice.[15] According to the classic-modern view of technology, instabilities exist in nature but ought to be excluded from technical systems. If instabilities arose, the traditional objective was to eliminate them. Controllability, based on predictability, expectability, and robustness, seemed feasible only when stability was guaranteed. Technology was equated with and defined by stability. Today, synthetic biologists—in line with computer scientists working in artificial intelligence and machine learning—are widening our understanding (and our concept) of technology by ascribing both stability *and* instability to technology. At the same time, it is still an open question whether the late-modern kind of technical system can be conclusively called "technology" or even whether it is a "technically possible technology" at all—to paraphrase the sociologist and system theorist Niklas Luhmann (2003, 100f). It can convincingly be argued that traditional "rational design" approaches in engineering and technology, which are typically based on assumptions of stability, have their limitations in the late-modern field of technology (cp. Giese et al. 2013). Alfred Nordmann (2008b, 173) states from a critical angle: "No longer a means of controlling nature in order to protect, shield, or empower humans, technology dissolves into nature and becomes uncanny, incomprehensible, beyond perceptual and conceptual control." Whenever instabilities are involved, non-knowledge, uncertainties, and ignorance also prevail and, in principle, cannot be eliminated; problems with regard to monitoring and controlling emerge. Late-modern technical systems have a life of their own; instabilities render engineering (construction/design and monitoring/controlling) difficult (Kastenhofer and Schmidt 2011).

It is highly interesting that the ethics of Hans Jonas is well equipped to address and to assess this novel kind of technology. Jonas's new future-oriented imperative—"'Act so that the effects of your action are compatible with the permanence of genuine human life'" (Jonas 1984, 11)—is much informed by his general reflection on the ambivalence encountered in the advancement of science and technology and especially evident in the technoscientific core of emerging technologies. Jonas anticipated the characteristics and limits of "engineering biology" even back in the mid-eighties (Jonas 1987, 163).[16] In extension to Jonas's terminology, I use the term "late-modern technology" to underline that we are experiencing a qualitative change in what we now consider technology. Jonas diagnoses a historically new technoscientific era and perceives a radical "newness of biologically based technology" (ibid., 163).[17] He draws a dividing line between the classic engineering type of technology—including what he calls the "art of the engineer" or, synonymously, "engineering art"—and a biologically based type of technology. As Jonas argues, this new type of technology differs in a qualitative way from our common perception and understanding of what technology is or could be.

> In th[is latter …] case of dead substances, the constructor is the one and only actor with respect to a passive material [= classic-modern technology.] [In contrast, in the case of the] biological organism, activity meets activity: biological technology is collaborative; it is self-activity of an active [= living] 'material'.[18]
>
> (ibid., 165)

Jonas lists characteristics of this new type of technology:

- self-activity, processuality, and autonomy;
- irreversibility, time-dependency, and historicity (birth and death);
- complexity, evolution, and growth;
- individuality, non-experimentability, and obstacles regarding reproducibility;
- collaborativeness and interactive causation as a different kind of causality (ibid., 163ff).

Jonas argues that, since biologically based technology inevitably carries an internal activity, engineering "means releasing the bio-object into the stream of becoming in which the engineer and constructor is also drifting" (ibid., 168).[19] Looking at the present wave of synthetic biology, Jonas's anticipation and, in particular, his differentiation between "engineering art" and "biologically based engineering" are certainly very fascinating. However, Jonas does not take his distinctly phenomenological description any further. In consonance with the argumentation developed in this chapter, it is not the organismic alone that constitutes the central difference but also instability-based self-organization. Jonas's approach is much more fundamental than Jonas himself seems to have assumed. Closer examination of the new type of technology provides a further argument in support of the need for a heuristics of fear comprising the precautionary principle and the imperative of responsibility.

In searching for *an ethics for the technological age*, Jonas (1984) anticipates that our notion and understanding of "technology" seem to be changing. Let us, for a moment, like Jonas assume that a late-modern type of technology could, in principle, become technically feasible, applicable, and successful. We would then be faced with new challenges such as restrictions with regard to predictability or limited control—the flip side of self-organization. The fundamental properties of such a late-modern technology (evolution, growth, autonomy, and self-productivity) have the power to change the world we live in. Metaphorically speaking, those who dare to stimulate and induce instabilities are, at the same time, provoking a risky *dance on the razor's edge*. "Because engineered micro-organisms are self-replicating and capable of evolution," Jonathan B. Tucker and Raymond A. Zilinskas (2006) argue, "they belong in a different risk category than toxic chemicals or radioactive materials." In fact, this objection already applies to a number of classic substances in biotechnology. But the related challenges in the realm of synthetic biology seem to go much deeper and could be regarded as more pressing. Notably, the extent to which the *principle of similarity and resemblance*, which constitutes the backbone of any risk assessment, is applicable to the substances and tissues used in synthetic biology remains open to debate. This principle is based on the assumption that if a new (bio-)system has some similarity to one that is known, the new system will behave in a similar way as the well-known one and exhibit essentially similar properties. But many self-organizing bio-systems are not all that similar, owing to their intrinsic instability and the production of novel features, functionalities, or substances. How, then, are they to be compared to other, well-known bio-systems? Such questions are ethically relevant, Jonas argues; they challenge the feasibility of an assessment and, consequently, also of ethics.

According to Jonas, it is of ethical relevance that non-knowledge, ignorance, and uncertainty are co-produced with the productiveness of the late-modern technical systems—that is a central point made by Jonas with his *heuristics of fear*. Non-knowledge and uncertainty are by-products and do not simply emerge in the societal context of diffusion, use, and consumption. Instability-based technology takes on a life of its own. Jean-Pierre Dupuy (2004, 10) is citing Jonas when he argues, "The novel kind of uncertainty that is brought about by those new technologies [...] is intimately linked with their being able to set off complex phenomena in the Neumannian sense" (cp. Dupuy and Grinbaum 2006, 289).[20] Because of the "unpredictable behavior [...] engineers will not know how to make [... these] machines until they actually start building them" (Dupuy 2004, 18). In a similar tenor, scholars from Prigogine's Brussels school of complexity have raised concerns regarding control options: We have "focused on designing and implementing artificial self-organizing systems in order to fulfill particular functions. Such systems have several advantages. [... However,] [d]isadvantages are limited predictability and difficulty of control" (Heylighen 2002).[21] The disadvantages become obvious when we consider the new and unknowable, instability-based risks and the "unknown unknowns." This thinking concurs with what Alfred Nordmann perceived as a "limit [that] could [...] be reached where engineering seeks to exploit surprising properties that arise from natural processes of self-organization" (Nordmann 2008b, 175). We are on the way to "surrender[ing] control to pervasive technical systems" (ibid., 182).

Summary and prospect

One might raise doubts as to whether well-established concepts of TA can address and assess the novel type of technology. According to Dupuy and Grinbaum (2006, 293), "none of these [TA] tools is appropriate for tackling the situation we are facing now." What Dupuy and Grinbaum are expressing is certainly true of classic TA approaches. However, as this chapter sketches, more recent directions in TA such as ProTA offer prospects to enable an early assessment. ProTA analyses the technoscientific core in detail. It is particularly relevant when it is a case of inquiring into alternatives (a) *within* or (b) *to* the technoscientific core itself and, based on this, searching for new or different directions in science, technology, and innovation policy (Schmidt 2016).

A central question emerges in this context (ad a): Can we identify research and development trajectories of synthetic biology that are aimed at designing bio-systems having internal safety features—for example, cell-free systems that share certain positive properties or desired functionalities with cell-based systems but are essentially less fraught with instability and therefore not capable of strong forms of self-organization?[22] Other questions address a more positive direction: Do certain subfields of synthetic biology carry realistic potential to meet the requirements for sustainable development?

In addition (ad b), a key issue on a much more fundamental and certainly more pressing level is whether our late-modern society should really foster and facilitate a "late-modern technology"—a technology that is inherently unstable and linked with the ability to set off self-organizing, complex, and autonomous dynamics.[23] Late-modern technology differs from the classic-modern type of technology with regard to three main categories of characteristics.

First, *phenomenological characteristics*: Late-modern technical systems are based on self-organization. They appear to be un-technical and non-artificial. They show autonomous behaviour and agency properties: Signs and signals, tracks and traces of technology are no longer visible. Culturally established borders are becoming blurred. This universal trend is leading towards a *phenomenological naturalization of technology*. Second, *nomological or ontological characteristics*: The nomological core of late-modern technology is instability—as a necessary condition for self-organization. Instabilities are intentionally built into the technical systems and their material structures. Here, we can perceive a trend that could be called *nomological naturalization of nature*. Third, *methodological, epistemological, and action-theoretical characteristics*: Late-modern technology is different from other types of technology in that certain criteria are absent. A late-modern technical system is hardly (a) separable from its environment and from the context of application; it lacks (b) reproducibility, (c) predictability, and (d) testability/describability; it gives rise to limitations with regard to (e) constructing and creating; and it eludes (f) monitoring and controlling.

Therefore, this kind of technology has, or if realized to its full extent would have, a life of its own. It could be regarded as a "naturalized technology" (Nordmann 2008b), denoting a *phenomenological* as well as a *nomological naturalization of technology*. Whether late-modern technology can be conclusively called "technology" and whether it is "as a technical system technically possible at all" remain open to debate (Luhmann 2003, 100).[24] Nevertheless, technical systems, devices, things, and objects based on instabilities and showing self-organizing phenomena are beginning to populate our life-world. From an ethical perspective, we need to address this instability-based, late-modern type of technology and undertake the task of developing procedures either to restrict and contain or to shape and deal with it.

Hans Jonas was precursory in this respect (Jonas 1987). His future-oriented ethics might serve as a fundament for a further assessment of synthetic biology. The anti-utopian precautionary principle—with its recognition of an objective indeterminacy of real futures and the limits of knowledge— constitutes a conservativism appreciating the "responsibility for existence." Jonas already anticipated the ethical challenges of this novel kind of technology back in the late 1970s. ProTA, in alignment with Jonas's ethics, could offer an interdisciplinary, critical-reflexive approach that enables us to analyse and assess the technoscientific core of this new wave of emerging technologies. From Jonas, we can learn that the central criterion for an ethical assessment of an emerging technoscientific wave is—to paraphrase Kant—that the condition for the possibility of TA and ethics has to be guaranteed. This possibility seems to be challenged in the field of advanced synthetic biology. A novel concept of this kind has been explicitly developed by Christoph Hubig (2015).

In essence, ProTA can be viewed as a paradigm of a critical-reflexive interdisciplinary practice—it is instrumentalist on a deeper and more fundamental level than what has been labelled instrumentalist-strategic interdisciplinarity. As such, it is an extension of well-established TA concepts but does not replace them. David Collingridge's (1980, 16) central questions "How can we get the technology we want [...], and how can we avoid technologies which we do not want to have?" could be reworded as follows: How can scientists and societal actors conceptualize, understand, and shape the technosciences and technoscientific knowledge in the way we want during the early phases of research and development processes?

ProTA advances an anticipatory approach to deal with these urgent challenges. Since it puts a critical and reflexive mindset at the very centre of technoscience-based knowledge production, it can be deemed to truly epitomize the concept of critical-reflexive interdisciplinarity.

Notes

1 Besides Prospective Technology Assessment (as first conceptualized in Liebert and Schmidt 2010), there are cognate concepts with similar perspectives, such as vision assessment (Grin and Grunwald 2000), real-time TA (Guston and Sarewitz 2002), constructive TA (Schot and Rip 1996), technology characterization (Gleich 2004), hermeneutical TA (Grunwald 2016), science assessment (Gill 1994), early-stage technology analysis (Zweck 2002), and, more generally, innovation and technology analysis; see also the introduction to TA in general: Grunwald (2019).

2 See, for instance, Fagerberg et al. (2005).

3 This fact is condensed in the diagnosis of the regime of technoscience, see Chapter 4 of this book.

4 See Chapter 6 in this book.

5 In applying the four orientations, ProTA complements the broad variety of existing TA (and related) studies on synthetic biology; to mention just a few: European Commission (2005), Miller and Selgelid (2006), Vriend (2006), Royal Academy of Engineering (2009), European Technology Assessment Group (2009), Schmidt (2009), and Giese et al. (2015).

6 On the one hand, "synthetic biology" seems to be a fairly young term. It was (re-)introduced and presented by Eric Kool in 2000 at the annual meeting of the American Chemical Society. Since then, the term has enjoyed a remarkable career and general circulation in the scientific communities as well as in science, technology, and innovation politics. On the other hand, the notion of synthetic biology emerged about 100 years ago—but it was rarely mentioned until the year 2000.

7 The European Technology Assessment Group (2009, 14) uses the term "paradigm" and states that synthetic biology can be considered a "new research paradigm." See also Schmidt and de Lorenzo (2012).

8 Nersessian and Patton (2009) and Nersessian (2012) have investigated the role of engineering concepts in biology, focusing in particular on the process of concept formation, sense making, and model-based reasoning.

9 Similar expressions can be found in Nolfi and Floreano (2000) and Schwille (2011).

10 The European Technology Assessment Group (2009, 25) goes on to stress: "Central in their ideas is the concept of self-regulation, self-organization and feedback as essential characteristics of cognitive systems since continuous adaption to the environment is the only way for living systems to survive."

11 My translation of the German version (J.C.S.).

12 My translation (J.C.S.). The term "structural science" is exemplified in Schmidt (2008a).

13 See, for example, Schmidt (2008a, 2011a, 2015).

14 See the previous chapter.

15 One could say, in a more provocative manner, that the *more* late-modern societies, facilitated by (the ideals of) synthetic biologists, seem to control the material world, the *more* they lose their ability to control it. A control dialectic emerges, as shown in Kastenhofer and Schmidt (2011) and Schmidt (2015a).

16 Jonas did so explicitly in a book chapter ("Lasst uns einen Menschen klonieren: Von der Eugenik zur Gentechnologie," in "Technik, Medizin und Ethik," Jonas 1987) published in German only.

17 My translation (J.C.S.).

18 My translation (J.C.S.). The new "collaborative kind of technology" seems to be closer to humans and their actions and self-perceptions; it is not alien to humans like the mechanical type of technology of classic-modern engineering. From the same perspective, and a few decades earlier in the late 1950s, Ernst Bloch coined the term "alliance technology" to underline the difference between mechanical-engineering and biology-based technology (Bloch 1995). According to Bloch, we may apply that term to a technology based on self-organization. Today, we need to go beyond the thinking of Bloch and consider also the ambivalence of this type of technology.

19 My translation (J.C.S.).

20 The famous physicist Richard Feynman is quoted as saying: "What I cannot create, I do not understand" (cp. Schwille and Diez 2009, 223).

21 In line with these concerns, Joy (2000) warns about the dangers of the well-known and highly disputed dystopia: the "gray goo."

22 Questions of this kind are posed by Gleich et al. (2012), Marliere (2009, 77f), and Schmidt and de Lorenzo (2013, 2201f).

23 This is in line with Nordmann's concerns as to whether we can cope with this kind of technology. His objections are far-reaching: "This is a critique no longer of what we do to nature in the name of social and economic control. Instead it is a critique of what we do to ourselves as we surrender control to pervasive technical systems" (Nordmann 2008b, 182).

24 My translation (J.C.S.).

References

Abarbanel, H. D., Brown, R., Sidorowich, J., Tsimring, L.S. (1993). The Analysis of Observed Chaotic Data in Physical Systems. *Reviews of Modern Physics* 65, pp. 1331–1392.

Achterhuis, H. (ed.) (2001). *American Philosophy of Technology: The Empirical Turn.* Bloomington, IN: Indiana University Press.

Action Group on Erosion, Technology and Concentration. (2007). *Extreme Genetic Engineering. An Introduction to Synthetic Biology.* http://www.etcgroup.org/en/materials/publications.html?pub_id=602. Accessed 23 May 2020.

Adorno, T.W. (1969). *Stichworte. Kritische Modelle.* Frankfurt: Suhrkamp.

Andronov, A., Pontryagin, L. (1937). Systèmes Grossiers. *Doklady Akademii Nauk (Doklady) SSSR* 14, pp. 247–251.

Andronov, A., Witt, A.A., Chaikin, S.E. (1965/1969). *Theorie der Schwingungen.* Part I/II. Berlin: Akademie.

Apel, K.-O. (1988). *Diskurs und Verantwortung.* Frankfurt: Suhrkamp.

Aubin, D., Dalmedico, A.D. (2002). Writing the History of Dynamical Systems and Chaos: Longue Durée and Revolution, Disciplines and Cultures. *Historia Mathematica* 29, pp. 273–339.

Aurell, E., Boffetta, G., Crisanti, A., Paladin, G, Vulpiani, A. (1997). Predictability in the Large: An Extension of the Concept of Lyapunov Exponent. *Journal of Physics A* 30, pp. 1–26.

Bacon, F. (1966). *The Advancement of Learning and New Atlantis* (1605). London: Oxford University Press.

Bacon, F. (2004). *The Instauratio magna Part II: Novum organon and Associated Texts* (1620), ed. G. Rees and M. Wakely. New York: Clarendon Press.

Baird, D. (2004). *Thing Knowledge. A Philosophy of Scientific Instruments.* London: University of California Press.

Baird, D., Nordmann, A., Schummer, J. (eds.) (2004). *Discovering the Nanoscale.* Amsterdam: IOS.

Baird, D., Shew, A. (2004). Probing the History of Scanning Tunneling Microscopy. In: Baird, D. et al. (eds.). *Discovering the Nanoscale.* Amsterdam: IOS, pp. 145–156.

Balsinger, P. (1999). Disziplingeschichtsschreibung und Interdisziplinarität. In: Thiel, C., Peckhaus, V. (eds.). *Disziplinen im Kontext. Perspektiven der Disziplingeschichtsschreibung.* München: Fink, pp. 232–242.

Balsinger, P. (2005). *Transdisziplinarität. Systematisch-vergleichende Untersuchung disziplinenübergreifender Wissenschaftspraxis.* München: Fink.

Bammé, A. (2004). *Science Wars. Von der akademischen zur postakademischen Wissenschaft.* Frankfurt: Campus.

Bammer, G. (2013). *Disciplining Interdisciplinarity: Integration and Implementation Sciences for Researching Complex Real-World Problems*. Canberra, Australia: Australian National University Press.

Batterman, R. (2002). *The Devil in the Detail. Asymptotic Reasoning in Explanation, Reduction and Emergence*. Oxford: Oxford University Press.

Baudrillard, J. (2003). *The Transparency of Evil. Essays on Extreme Phenomena*. London/New York: Verso.

Bechmann, G., Frederichs, G. (1996). Problemorientierte Forschung. Zwischen Politik und Wissenschaft. In: Bechmann, G. (ed.). *Praxisfelder der Technikfolgenforschung*. Frankfurt: Campus, pp. 11–37.

Beck, U. (1992). *Risk Society. Towards a New Modernity*. London: Sage Publications.

Beck, U. (2007). *Weltrisikogesellschaft. Auf der Suche nach der verlorenen Sicherheit*. Frankfurt: Suhrkamp.

Beck, U., Lau, C. (eds.) (2004). *Entgrenzung und Entscheidung: Was ist neu an der Theorie reflexiver Modernisierung?* Frankfurt: Suhrkamp.

Becker, E. (2002). Transformation of Social and Ecological Issues into Transdisciplinary Research. In: UNESCO (ed.). *Knowledge for Sustainable Development. An Insight into the Encyclopedia of Life Support Systems*, vol 3. UNESCO, Paris, pp. 949–963.

Becker, E., Jahn, T. (eds.) (2006). *Soziale Ökologie. Grundzüge einer Wissenschaft von den gesellschaftlichen Naturverhältnissen*. Frankfurt: Campus.

Benyus, J.M. (2002). *Biomimicry: Innovation Inspired by Nature*. New York: HarperCollins.

Bergé, P., Pomeau, Y., Vidal, C. (1984). *Order within Chaos. Towards a Deterministic Approach to Turbulence*. New York: Wiley.

Bergmann, M. et al. (2012). *Methods for Transdisciplinary Research. A primer for practice*. Frankfurt/New York: Campus.

Bernstein, J.H. (2015). Transdisciplinarity: A Review of Its Origins, Development, and Current Issues. *Journal of Research Practice* 11(1), pp. 1–20 (article R1).

Bijker, W.E., Law, J. (eds.) (1994). *Shaping Technology / Building Society. Studies in Sociotechnical Change* (1992). Cambridge, MA: MIT Press.

Birkhoff, G.D. (1927). *Dynamical Systems*. New York: American Mathematical Society.

Bloch, E. (1995). *The Principle of Hope (1959)*. Cambridge, MA: MIT Press.

Bloor, D. (1999). Anti-Latour. *Studies in History and Philosophy of Science* 30(1), pp. 81–122.

Bogner, A., Kastenhofer, K., Torgersen, H. (eds.) (2010). *Inter- und Transdisziplinarität im Wandel? Neue Perspektiven auf problemorientierte Forschung und Politikberatung*. Baden-Baden: Nomos.

Böhler, D. (ed.) (1994). *Ethik für die Zukunft. Im Diskurs mit Hans Jonas*. München: Beck.

Böhler, D., Bongardt, M., Burckhart, H., Wiese, C. Zimmerli, W.C. (eds.) (2009ff). *Kritische Hans Jonas Werksausgabe*. Freiburg: Rombach.

Böhme, G. (1993). *Am Ende des Baconschen Zeitalters*. Frankfurt: Suhrkamp.

Böhme, G. (1999). *Bios – Ethos. Über ethikrelevantes Naturwissen*. Bremen: Manholt.

Böhme, G. et al. (eds.) (1983). *Finalization in Science. The Social Orientation of Scientific Progress*. Dordrecht: Reidel.

Böhme, G., Stehr, N. (1986). *The Knowledge Society*. Dordrecht: Reidel.

Böhme, G, van den Daele, W. (1977). Erfahrung als Programm – Über Strukturen vorparadigmatischer Wissenschaft. In: Böhme, G., van den Daele, W., Krohn, W. (eds.). *Experimentelle Philosophie*. Frankfurt: Suhrkamp, pp. 183–236.

Briggle, A. (2010). *A Rich Bioethics: Public Policy, Biotechnology, and the Kass Council*. Notre Dame, IN: University of Notre Dame Press.

Brown, R., Chua, L.O. (1996). Clarifying Chaos: Examples and Counterexamples. *International Journal of Bifurcation and Chaos* 6(2), pp. 219–249.

Bunge, M. (1987). *Kausalität. Geschichte und Probleme*. Tübingen: Mohr.

Bush, V. (1945). *Science: The Endless Frontier. A Report to the President on a Program for Postwar Scientific Research*. Washington, DC: United States Government Printing Office. http://www.nsf.gov/od/lpa/nsf50/vbush1945.htm. Accessed 10 August 2020.

Carnap, R. (1928). *Der logische Aufbau der Welt*. Berlin: Weltkreis.

Carrier, M. (2001). Business as Usual: On the Prospect of Normality in Scientific Research. In: Decker, M. (ed.). *Interdisciplinarity in Technology Assessment. Implementation and Its Chances and Limits*. Berlin: Springer, pp. 25–31.

Carrier, M. (2011). 'Knowledge Is Power', Or How to Capture the Relationship between Science and Technoscience. In: Nordmann, A. et al. (eds.). *Science Transformed? Debating Claims of an Epochal Break*. Pittsburgh, PA: University of Pittsburgh Press, pp. 43–53.

Carter, B. (1974). Large Number Coincidences and the Anthropic Principle in Cosmology. In: Longguir, M. (ed.). *Confrontation of Cosmological Theories with Observational Data*. Dordrecht: Reidel, pp. 291–298.

Cartwright, N. (1983). *How the Laws of Physics Lie*. Oxford: Oxford University Press.

Cartwright, N. (1994). *Nature's Capacities and Their Measurement*. Oxford: Oxford University Press.

Cartwright, N. (1999). *The Dappled World: A Study of the Boundaries of Science*. Cambridge, UK: Cambridge University Press.

Chaitin, G.J. (1971). Computational Complexity and Gödel's Incompleteness Theorem. *ACM SIGACT News* 9, pp. 11–12.

Chaitin, G.J. (1987). *Algorithmic Information Theory*. Cambridge, UK: Cambridge University Press.

Chandler, D. (2018). *Ontopolitics in the Anthropocene*. London/New York: Routledge.

Chubin, S., Porter, A.L., Rossini, F.A., Connolly, T. (eds.) (1986). *Interdisciplinary Analysis and Research. Theory and Practice of Problem-Focused Research and Development*. Airy, MD: Mt Lomond Publications.

Collingridge, D. (1980). *The Social Control of Technology*. New York: St. Martin's Press.

Collingridge, D., Reeve, C. (1986). *Science Speaks to Power: The Role of Experts in Policy Making*. London: Frances Pinter.

Comte, A. (2006). *Cours de Philosophie Positive (1830–1842)*. Paris: Bachelier.

Cozzens, S.E., Gieryn, T.F. (eds.) (1990). *Theories of Science in Society*. Bloomington, IN: Indiana University Press.

Crutchfield, J.P., Farmer, J.D., Packard, N.H., Shaw, R.S. (1986). Chaos. *Scientific American* 12, pp. 46–57.

Csete, M.E., Doyle, J.C. (2002). Reverse Engineering of Biological Complexity. *Science* 295, pp. 1664–1669.

Darrigol, O. (2008). *Worlds of Flow. A History of Hydrodynamics from the Bernoullis to Prandtl (2005)*. Oxford: Oxford University Press.

Daston, L., Galison, P. (2007). *Objectivity*. New York: Zone Books.

Decker, M. (ed.) (2001). *Interdisciplinarity in Technology Assessment. Implementation and Its Chances and Limits*. Berlin: Springer

Decker, M. (2004). The Role of Ethics in Interdisciplinary Technology Assessment. *Poiesis & Praxis 2(2–3)*, pp. 139–156.

Decker, M. (2010). Interdisziplinäre Wissensgenerierung in der TA—eine Prozessbeschreibung. In: Bogner, A., Kastenhofer, K., Torgersen, H. (eds.). *Inter- und Transdisziplinarität im Wandel? Neue Perspektiven auf problemorientierte Forschung und Politikberatung*. Baden-Baden: Nomos, pp. 145–165.

Decker, M., Grunwald, A. (2001). Rational Technology Assessment as Interdisciplinary Research. In: Decker, M. (ed.). *Interdisciplinarity in Technology Assessment. Implementation and Its Chances and Limits*. Berlin: Springer, pp. 33–60.

Deplazes, A., Huppenbauer, M. (2009). Synthetic Organisms and Living Machines: Positioning the Products of Synthetic Biology at the Borderline between Living and Nonliving Matter. *Systems and Synthetic Biology* 3, pp. 55–63.

Descartes, R. (1979). *Regeln zur Ausrichtung der Erkenntniskraft (1629)*. Hamburg: Meiner.

Devaney, R.L. (1987). *An Introduction to Chaotic Dynamical Systems*. Redwood City, CA: Addison Wesley.

Dewey, J. (1929). *The Quest for Certainty: A Study of the Relation of Knowledge and Action*. New York: Minton, Balch & Co.

DFG, acatech, Leopoldina (2009). *Synthetische Biologie. Stellungnahme*. Bonn: Wiley-VCH.

Dörner, D. (1995). *Die Logik des Mißlingens. Strategisches Denken in komplexen Situationen*. Hamburg: Rowohlt.

Drexler, K.E. (1990). *Engines of Creation: The Coming Era of Nanotechnology*. London/Oxford: Oxford University Press.

Drubin, D.A., Way, J.C., Silver, P.A. (2007). Designing Biological Systems. *Genes and Development* 21, pp. 242–254.

Duhem, P. (1991). *The Aim and Structure of Physical Theory (1906)*. Princeton, NJ: Princeton University Press.

Dupuy, J.-P. (2004). *Complexity and Uncertainty: A Prudential Approach to Nanotechnology*. In: European Commission - Health and Consumer Protection Directorate General (eds.): Nanotechnologies. Brüssel, pp. 78–94.

Dupuy, J.-P. (2009). The Precautionary Principle and Enlightened Doomsaying: Rational Choice before the Apocalypse. *Occasion: Interdisciplinary Studies in the Humanities* 1(1) (October 15). http://occasion.stanford.edu/node/28.

Dupuy, J.-P., Grinbaum, A. (2006). Living with Uncertainty: Toward the Ongoing Normative Assessment of Nanotechnology. In: Baird, D., Schummer, J. (eds.). *Nanotechnology Challenges: Implications of Philosophy, Ethics and Society*. Singapore: World Scientific, pp. 287–314.

Ebeling, W., Feistel, R. (1994). *Chaos und Kosmos. Prinzipien der Evolution*. Heidelberg/Berlin: Spektrum Akademischer Verlag.

Einstein, A. (1917). Kosmologische Betrachtungen zur allgemeinen Relativitätstheorie. *Sitzungsberichte der Königlich Preussischen Akademie der Wissenschaften, physikalisch-mathematische Klasse*, pp. 142–152.

Einstein, A., Podolsky, B., Rosen, N. (1935). Can Quantum-Mechanical Description of Physical Reality Be Considered Complete? *Physics Review* 47, pp. 777–780.

Endy, D. (2005). Foundations for Engineering Biology. *Nature* 438, pp. 449–453.

Etzkowitz, H., Leydesdorff, L. (eds.) (1997). Special Issue on Science Policy Dimensions of the Triple Helix of University–Industry–Government Relations. *Science and Public Policy* 24(1), pp. 2–52.

Etzkowitz, H., Leydesdorff, L. (1998). The Endless Transition: A "Triple Helix" of University-Industry-Government Relations. *Minerva* 36(3), pp. 271–288/203–208.

Euler, P. (1999). *Technologie und Urteilskraft*. Weinheim: Deutscher Studien Verlag.

European Commission (2005). *Synthetic Biology. Applying Engineering to Biology* (Report of a NEST High Level Expert Group EU 21796). Brussels: European Commission.

European Technology Assessment Group (2009). *Making a Perfect Life. Bioengineering in the 21st Century*. The Hague: Rathenau Institute.

Fagerberg, J. (2005). Innovation—A Guide to the Literature. In: Fagerberg, J., Mowery, D.C., Nelson, R.R. (eds.). *The Oxford Handbook of Innovation*. Oxford: Oxford University Press, pp. 1–26.

Fagerberg, J., Mowery, D.C., Nelson, R.R. (eds.) (2005). *The Oxford Handbook of Innovation*. Oxford: Oxford University Press.

Feenberg, A. (1991). *Critical Theory of Technology*. Oxford: Oxford University Press.

Feenberg, A. (2002). *Transforming Technology. A Critical Theory Revisited*. Oxford: Oxford University Press.

Feynman, R.E. (2003). There's Plenty of Room at the Bottom (1959). http://www.zyvex.com/nanotech/feynman.html. Accessed 12 May 2014.

Fleck, L. (1979). *Genesis and Development of a Scientific Fact (1935)*. Chicago: University of Chicago Press.

Forman, P. (2012). On the Historical Forms of Knowledge, Production and Curation: Modernity Entailed Disciplinarity, Postmodernity Entails Antidisciplinarity. *Osiris* 27, pp. 56–97.

Frankena, W.K. (1973). *Ethics*. Englewood Cliffs, CA: Prentice-Hall.

Frodeman, R. (2010). Introduction. In: Frodeman, R., Klein, J.T., Mitcham, C. (eds.). *The Oxford Handbook of Interdisciplinarity*. Oxford: Oxford University Press, pp. xxix–xxxix.

Frodeman, R. (2014). *Sustainable Knowledge. A Theory of Interdisciplinarity*. New York/London: Palgrave MacMillan.

Frodeman, R., Briggle, A. (2016). *Socrates Tenured. The Institutions of 21st-Century Philosophy*. London/New York: Rowland & Littlefield.

Frodeman, R., Mitcham, C. (eds.) (2004). New Directions in the Philosophy of Science. Toward a Philosophy of Science Policy, Special Issue. *Philosophy Today* 48(5).

Frodeman, R., Mitcham, C. (2007). New Directions in Interdisciplinarity: Broad, Deep, and Critical. *Bulletin of Science, Technology and Society* 27(6), pp. 506–514.

Frodeman, R., Klein, J.T., Mitcham, C. (eds.) (2010). *The Oxford Handbook of Interdisciplinarity*. Oxford: Oxford University Press.

Fuller, S. (2002). *Social Epistemology (1988)*. Bloomington, IN: Indiana University Press.

Fuller, S. (2010). Deviant Interdisciplinarity. In: Frodeman, R., Klein, J.T., Mitcham, C. (eds.) (2010). *The Oxford Handbook of Interdisciplinarity*. Oxford: Oxford University Press, pp. 50–64.

Fuller, S. (2017). The Military-Industrial Route to Interdisciplinarity. In: Frodeman, R., Klein, J.T., Pacheco, R.C.D. (eds.) (2017). *The Oxford Handbook of Interdisciplinarity* (2nd ed.). Oxford: Oxford University Press, pp. 53–67.

Funtowicz, S.O., Ravetz, J.R. (1993). Science for the Post-Normal Age. *Futures* 9, pp. 739–755.

Galison, P. (1996). Computer Simulations and the Trading Zone. In: Galison, P., Stump, D.J. (eds.). *The Disunity of Science. Boundaries, Contexts, and Power*. Stanford, CA: Stanford University Press, pp. 118–157.

Galison, P., Stump, D.J. (eds.) (1996). *The Disunity of Science. Boundaries, Contexts, and Power.* Stanford, CA: Stanford University Press.

Gehring, P. (2006). *Was ist Biomacht? Vom zweifelhaften Mehrwert des Lebens.* Frankfurt: Campus.

Gethmann, C.F. (1999). Rationale Technikfolgenbeurteilung. In: Grunwald, A. (ed.). *Rationale Technikfolgenbeurteilung. Konzepte und methodische Grundlagen.* Berlin: Springer, pp. 1–10.

Gethmann, C.F., Carrier, M. et al. (2015). *Interdisciplinary Research and Transdisciplinary Validity Claims.* Berlin/Heidelberg: Springer.

Gibbons, M. et al. (1994). *The New Production of Knowledge.* London: Sage.

Gibson, D.G., Glass, J. I., Lartigue, C., Noskov, V.N., Chuang, R.Y., Algire, M.A., et al. (2010). Creation of a Bacterial Cell Controlled by a Chemically Synthesized Genome. *Science* 329 (5987), pp. 52–56.

Gierer, A. (1981). Physik der biologischen Gestaltbildung. *Naturwissenschaften* 68, pp. 245–251.

Gieryn, T.F. (1983). Boundary-Work and the Demarcation of Science from Non-Science: Strains and the Interests in Professional Ideologies of Scientists. *American Sociological Review* 48, pp. 781–795.

Giese, B. et al. (2013). Rational Engineering Principles in Synthetic Biology: A Framework for Quantitative Analysis and an Initial Assessment. *Biological Theory* 8, pp. 324–333.

Giese, B. et al. (eds.) (2015). *Synthetic Biology. Character and Impact.* Berlin/New York: Springer.

Gill, B. (1994). Folgenerkenntnis. Science Assessment als Selbstreflexion der Wissenschaft. *Soziale Welt* 45, pp. 430–453.

Gleich, A.V. (2004). Technikcharakterisierung als Ansatz einer vorsorgeorientierten prospektiven Innovations- und Technikanalyse. In: Bora, A. et al. (eds.). *Technik in einer fragilen Welt. Die Rolle der Technikfolgenabschätzung.* Berlin: Sigma, pp. 229–244.

Gleich, A.v., Giese, B., Königstein, S., Schmidt, J.C. (2012). *Synthetische Biologie: Revolution oder Evolution? Definition, Charakterisierung und Entwicklungsperspektiven der Synthetischen Biologie mit Fokus auf den damit verbundenen Chancen und Risiken.* Bremen: afortec.

Gleick, J. (1987). *Chaos: Making a New Science.* New York: Viking Penguin.

Goethe, J. W. V. (2006). *Theory of Colors* (1st ed., ed. John Murray. London, 1840). Mineola/New York: Dover Publication.

Gorman, M. E. (ed.). (2010). *Trading Zones and Interactional Expertise.* Cambridge, MA: MIT Press.

Graff, H.J. (2015). *Undisciplining Knowledge: Interdisciplinarity in the Twentieth Century.* Baltimore, MD: Johns Hopkins University Press.

Grin, J., Grunwald, A. (eds.) (2000). *Vision Assessment. Shaping Technology in 21st Century Society.* Berlin/New York: Springer.

Grunwald, A. (1996). Ethik der Technik. Systematisierung und Kritik vorliegender Entwürfe. *Ethik und Sozialwissenschaften* 2/3, pp. 191–204.

Grunwald, A. (ed.) (1999). *Rationale Technikfolgenbeurteilung. Konzepte und methodische Grundlagen.* Berlin: Springer.

Grunwald, A. (2002). *Technikfolgenabschätzung. Eine Einführung.* Berlin: Sigma.

Grunwald, A. (2004). *Relevanz und Risiko. Zum Qualitätsmanagement integrativer Forschung* (Working Paper). Karlsruhe: KIT-Press.

Grunwald, A. (2015). Transformative Wissenschaft – eine neue Ordnung im Wissenschaftsbetrieb? *Gaia* 24(1), pp. 17–20.

Grunwald, A. (2016). *The Hermeneutic Side of Responsible Research and Innovation*. New York: Wiley.

Grunwald, A. (2019). *Technology Assessment in Practice and Theory*. London/New York: Routledge.

Grunwald, A., Schmidt, J.C. (eds.) (2005). Method(olog)ische Fragen der Inter- und Transdisziplinarität. Wege zu einer praxisstützenden Interdisziplinaritätsforschung. *Technikfolgenabschätzung – Theorie und Praxis* 14(2).

Guckenheimer, J., Holmes, P. (1983). *Nonlinear Oscillations, Dynamical Systems, and Bifurcations of Vector Fields*. New York: Springer.

Gusdorf, G. (1977). Past, Present, and Future in Interdisciplinary Research. *International Social Science Journal* 29, pp. 580–600.

Guston, D.H., Sarewitz, D. (2002). Real-time Technology Assessment. *Technology in Society* 24(1/2), pp. 93–109.

Habermas, J. (1970). *Toward a Rational Society*. Boston: Beacon Press.

Habermas, J. (1971). *Knowledge and Human Interest*. Boston: Beacon Press.

Habermas, J. (1984). *Theory of Communicative Action*. Boston: Beacon Press

Habermas, J. (1992). *Postmetaphysical Thinking. Philosophical Essays*. Cambridge, MA: MIT Press.

Habermas, J. (1993). *Justification and Application. Remarks on Discourse Ethics*. Cambridge, MA: MIT Press.

Habermas, J. (2002). *Die Zukunft der menschlichen Natur. Auf dem Weg zu einer liberalen Eugenik?* Frankfurt: Suhrkamp.

Hackett, E.J. et al. (eds.) (2008). *The Handbook of Science and Technology Studies*. 3rd ed. Cambridge, MA: MIT Press.

Hacking, I. (1983). *Representing and Intervening. Introductory Topics in the Philosophy of Natural Science*. New York: Cambridge University Press.

Hacking, I. (1999). *The Social Construction of What?* Cambridge, MA: Harvard University Press.

Haken, H. (ed.) (1980). *Dynamics of Synergetic Systems*. Berlin: Springer.

Hampe, M. (2006). *Erkenntnis und Praxis. Zur Philosophie des Pragmatismus*. Frankfurt: Suhrkamp.

Haraway, D. (1991). *Simians, Cyborgs, and Women: The Reinvention of Nature*. New York: Routledge.

Haraway, D. (2003). Modest_witness@second_millennium. In: MacKenzie, D., Wajcman, J. (eds.). *The Social Shaping of Technology*. Berkshire: Open University Press, pp. 41–49.

Harris, C.E., Pritchard, M.S., and Rabins, M.J. (2005). *Engineering Ethics. Concepts and Cases* (3rd ed.). Belmont, CA: Thomson Wadsworth.

Harremoes, P. et al. (eds.) (2001). *Late Lessons from Early Warnings. Precautionary Principle 1896–2000* (Issue Report No. 22). Copenhagen: European Environment Agency.

Harrell, M., Glymour, C. (2002). Confirmation and Chaos. *Philosophy of Science* 69, pp. 256–265.

Hartmann, S., Frigg, R. (2006). Models in Science. In: Zalta, E.N. (ed.). *Stanford Encyclopedia of Philosophy*. http://plato.stanford.edu/entries/models-science. Accessed 23 May 2020.

Hastedt, H. (1994). *Aufklärung und Technik*. Frankfurt: Suhrkamp.

Heidegger, M. (1977). *The Question Concerning Technology and Other Essays* (1962). New York/London: Garland Publishing.

Heidegger, M. (1986). *Nietzsches metaphysische Grundstellung im abendländischen Denken. Die ewige Wiederkehr des Gleichen* (Gesamtausgabe 44). Frankfurt: Klostermann.

Hempel, C.G. (1965). *Aspects of Scientific Explanation*. New York: Free Press.

Hentig, H.v. (1972). *Magier oder Magister? Über die Einheit der Wissenschaft im Verständigungsprozess*. Stuttgart: Klett.

Hertz, H. (1963). *Die Prinzipien der Mechanik. In neuem Zusammenhang dargestellt* (1894). Darmstadt: Wiss. Buchgesellschaft.

Heussler, H. (1889). *Francis Bacon und seine geschichtliche Stellung. Ein analytischer Versuch*. Breslau: Koebner.

Heylighen, F. (2002). *The Science of Self-Organization and Adaptivity*. Working Paper. Brussels: Center Leo Apostel/Free University of Brussels.

Hill, B. (1998). *Erfinden mit der Natur*. Aachen: Shaker Verlag.

Hirsch, M. (1984). The Dynamical Systems Approach to Differential Equations. *Bulletin of the American Mathematical Society* 11, pp. 1–64.

Hirsch Hadorn, G. (2000). *Umwelt, Natur und Moral*. Freiburg: Alber.

Hirsch Hadorn, G. et al. (eds.) (2008). *Handbook of Transdisciplinary Research*. New York/Dordrecht: Springer.

Höffe, O. (1993). *Moral als Preis der Moderne. Ein Versuch über Wissenschaft, Technik und Umwelt*. Frankfurt: Suhrkamp.

Holbrook, B. (2010). Peer Review. In: Frodeman, R., Klein, J.T., Mitcham, C. (eds.). *The Oxford Handbook of Interdisciplinarity*. Oxford: Oxford University Press, pp. 319–332.

Holbrook, B. (2013). What Is Interdisciplinary Communication? Reflections on the Very Idea of Disciplinary Integration. *Synthese* 190, pp. 865–1879.

Holmes, P. (2005). Ninety Plus Thirty Years of Nonlinear Dynamics: More Is Different and Less Is More. *International Journal of Bifurcation and Chaos* 15(9), pp. 2703–2716.

Hooker, C.A. (2004). Asymptotics, Reduction and Emergence. *British Journal for the Philosophy of Science* 55(3), pp. 435–479.

Horkheimer, M. (1972). *Critical Theory: Selected Essays* (translated by M.J. O'Connell and others). Toronto: Herder and Herder.

Horkheimer, M., Adorno, T.W. (1972). *Dialectic of Enlightenment* (translated by J. Cumming). New York: Herder and Herder.

Hösle, V. (1994). Ontologie und Ethik bei Hans Jonas. In: Böhler, D. (ed.). *Ethik für die Zukunft. Im Diskurs mit Hans Jonas*. München: Beck, pp. 105–125.

Hottois, G. (1984). *Le signe et la technique*. Paris: Aubier.

Hübenthal, U. (1991). *Interdisziplinäres Denken*. Stuttgart: Hirzel.

Hubig, C. (1995).*Technik- und Wissenschaftsethik*. Berlin: Springer.

Hubig, C. (2001). Interdisziplinarität und Abduktionenwirrwarr. Konkurrenz der Kompetenzen und Möglichkeiten einer Ordnung. In: Gottschalk-Mazouz, N., Mazouz, N. (eds.). *Nachhaltigkeit und globaler Wandel. Integrative Forschung zwischen Narrativität und Unsicherheit*. Frankfurt: Campus, pp. 319–340.

Hubig, C. (2006). *Die Kunst des Möglichen I. Technikphilosophie als Reflexion der Medialität*. Bielefeld: Transcript.

Hubig, C. (2015). *Die Kunst des Möglichen III. Macht der Technik*. Bielefeld: Transcript.

Hubig, C., Harrach, S. (2014). Transklassische Technik und Autonomie. In: Kaminski, A., Gelhard, A. (eds.). *Zur Philosophie informeller Technisierung*. Darmstadt: WGB, pp. 41–57.

Hume, D. (1990). *Enquiry Concerning Human Understanding and Concerning the Principles of Morals* (1748/1777). Oxford: Clarendon Press.

Hummel, D. et al. (2017). Social Ecology as Critical, Transdisciplinary Science—Conceptualizing, Analyzing and Shaping Societal Relations to Nature. *Sustainability 9*, pp. 1050–1070.

Hund, F. (1972). *Geschichte der physikalischen Begriffe*. Mannheim: BI.

Husserl, E. (1950). *Gesammelte Werke (Husserliana)*. Den Haag: Martinus Nijhoff.

Ihde, D. (1991). *Instrumental Realism: The Interface between Philosophy of Science and Philosophy of Technology*. Indianapolis, IN: Indiana University Press.

Ihde, D. (2002). *Bodies in Technology*. Minneapolis, MN: University of Minnesota Press.

Ihde, D., Selinger, E. (eds.) (2003). *Chasing Technoscience: Matrix for Materiality*. Indianapolis, IN: Indiana University Press.

Irwin, A. (1997). *Citizen Science: A Study of People, Expertise and Sustainable Development*. In: *Science, Technology and Human Values* 22(4), pp. 525–527.

Jacobs, J.A. (2013). *In Defense of Disciplines. Interdisciplinarity and Specialization in the Research University*. Chicago: University of Chicago Press.

Jacobs, J.A., Frickel, S. (2009). Interdisciplinarity: A Critical Assessment. *Annual Review of Sociology 35*, pp. 43–65.

Jaeger, J., Scheringer, M. (1998). Transdisziplinarität. Problemorientierung ohne Methodenzwang. *Gaia* 7(1), pp. 10–25.

Jaeger, J., Scheringer, M. (2018). Weshalb ist die Beteiligung von Akteuren nicht konstitutiv für transdisziplinäre Forschung. *Gaia* 27(4), pp. 345–347.

Jahn, T. (2013). Wissenschaft für eine nachhaltige Entwicklung braucht eine kritische Orientierung. *Gaia* 22(1), pp. 29–33.

Jahn, T., Bergmann, M., Keil, F. (2012). Transdisciplinarity: Between mainstreaming and marginalization. *Ecological Economics 79*, pp. 1–10.

James, W., (1977). *The Writings of William James* (here: Pragmatism and Other Writings, 1907), ed. J. McDermott. Chicago: University of Chicago Press.

Janich, P. (ed.) (1984). *Methodische Philosophie. Beiträge zum Begründungsproblem der exakten Wissenschaften in Auseinandersetzung mit Hugo Dingler*. Mannheim: Bibliographisches Institut.

Janich, P. (ed.) (1992). *Entwicklungen der methodischen Philosophie*. Frankfurt: Suhrkamp.

Janich, P. (1997). *Kleine Philosophie der Naturwissenschaften*. München: Beck.

Jantsch, E. (1970). Inter- and Transdisciplinary University: A Systems Approach to Education and Innovation. *Policy Sciences* 1(1), pp. 403–428.

Jantsch, E. (1972). Towards Interdisciplinarity and Transdisciplinarity in Education and Innovation. In: CERI (ed.). *Interdisciplinarity*. Paris: OECD, pp 97–121.

Jantsch, E. (1980). *The Self-Organizing Universe. Scientific and HumanImplication of the Emerging Paradigm of Evolution*. New York: Pergamon.

Jasanoff, S. et al. (eds.) (1994). *Handbook of Science and Technology Studies*. Thousand Oaks, CA: Sage.

Jonas, H. (1979). *Das Prinzip Verantwortung. Versuch einer Ethik für die technologische Zivilisation*. Frankfurt: Insel.

Jonas, H. (1984). *The Imperative of Responsibility. In Search of an Ethics for the Technological Age (1979)*. Chicago: University of Chicago Press.

Jonas, H. (1987). *Technik, Medizin und Ethik*. Frankfurt: Suhrkamp.

Jonas, H. (1993). *Dem bösen Ende näher*. Frankfurt: Suhrkamp.

Jonas, H. (1996). *Mortality and Morality. A Search for the Good after Auschwitz*. Evanston, IL: Northwestern University Press.

Jonas, H. (1997). *Das Prinzip Leben* (Engl. first ed. 1966; German version: Organismus und Freiheit. Ansätze zu einer philosophischen Biologie, 1973). Frankfurt: Suhrkamp.

Jonas, H. (2001). *The Phenomenon of Life: Toward a philosophical biology* (first ed. 1966). Evanston, IL: Northwestern University Press.

Jonas, H. (2009). Technology and Responsibility: Reflections on the New Task of Ethics. In: Winston, M.E., Edelbach, R.D. (eds.). *Society, Ethics and Technology* (4th ed.). Belmont, CA: Wadsworth, pp. 121–131.

Jones, R. (2004). *Soft Machines*. Oxford: Oxford University Press.

Joy, B. (2000). Why the Future Doesn't Need Us. Our Most Powerful 21st Century Technologies – Robotics, Genetic Engineering, and Nanotech – Are Threatening to Make Humans an Endangered Species. *Wired* 8(04), pp. 238–262.

Jungert, M. et al. (eds.) (2010). *Interdisziplinarität. Theorie, Praxis, Probleme.* Darmstadt: Wissenschaftliche Buchgesellschaft.

Kaku, M. (1998). *Visions. How Science Will Revolutionize the 21st Century.* New York: Anchor Books.

Karafyllis, N.C. (2007). Growth of Biofacts: The Real Thing or Metaphor? In: Heil, R., Kaminski, A., Stippack, M., Unger, A., Ziegler, M. (eds.). *Tensions and Convergences. Technological and Aesthetic (Trans-) Formations of Society.* Bielefeld: Transcript, pp. 141–152.

Kastenhofer, K. (2010). Zwischen schwacher und starker Interdisziplinarität: Sicherheitsforschung zu neuen Technologien. In: Bogner, A., Kastenhofer, K., Torgersen, H. (eds.). *Inter- und Transdisziplinarität im Wandel? Neue Perspektiven auf problemorientierte Forschung und Politikberatung.* Baden-Baden: Nomos, pp. 87–122.

Kastenhofer, K., Schmidt, J.C. (2011). On Intervention, Construction and Creation: Power and Knowledge in Technoscience and Late-Modern Technology. In: Zülsdorfer, T.B. et al. (eds.). *Quantum Engagements: Social Reflections of Nanoscience and Emerging Technologies.* Heidelberg: Akademische Verlagsgesellschaft, pp. 177–194.

Kates, R.W. et al. (2001). Sustainability Science. *Science* 292, pp. 641–642.

Kellert, S. (1994). *In the Wake of Chaos: Unpredictable Order in Dynamical Systems.* Chicago: University of Chicago Press.

Kincaid, H., Dupré, J., Wylie, A. (eds.) (2007). *Value-Free Science? Ideals and Illusions.* New York: Oxford University Press.

Klein, J.T. (1990). *Interdisciplinarity: History, Theory, and Practice.* Detroit, MI: Wayne State University.

Klein, J.T. (1996). *Crossing Boundaries: Knowledge, Disciplinarities, and Interdisciplinarities.* Charlottesville, VA: University Press of Virginia.

Klein, J.T. (2000). A Conceptual Vocabulary of Interdisciplinary Science. In: Weingart. P., Stehr, N. (eds.). *Practising Interdisciplinarity.* Toronto: University of Toronto Press, pp. 3–24.

Klein, J.T. (2010). A Taxonomy of Interdisciplinarity. In: Frodeman, R. et al. (eds.). *The Oxford Handbook of Interdisciplinarity.* Oxford: Oxford University Press, pp. 15–49.

Klein, J.T. et al. (eds.) (2001). *Transdisciplinarity: Joint Problem Solving Among Science, Technology, and Society.* Basel/Boston/Berlin: Birkhauser.

Kline, S.J. (1995). *Conceptual Foundations of Multidisciplinary Thinking.* Stanford, CA: Stanford University Press.

Knorr Cetina, K. (1999). *Epistemic Cultures: How the Sciences Make Knowledge.* Cambridge, MA: Harvard University Press.

Kocka, J. (ed.) (1987). *Interdisziplinarität. Praxis, Herausforderung, Ideologie.* Frankfurt: Suhrkamp.

Kockelmans, J.J. (ed.) (1979). *Interdisciplinarity and Higher Education.* University Park, PA: The Pennsylvania State University Press.

Köchy, K. (2011). Konstruktion von Leben? Herstellungsideale und Machbarkeitsgrenzen in der Synthetischen Biologie. In Gerhardt, V. et al. (eds.). *Evolution. Theorie, Formen und Konsequenzen eines Paradigmas in Natur, Technik und Kultur.* Berlin: Akademie Verlag, pp. 233–242.

Kötter, R., Balsiger, P.W. (1999). Interdisciplinarity and Transdisciplinarity: A Constant Challenge to the Sciences. *Integrative Studies* 17, pp. 87–120.

Krebs, A. (1997). *Naturethik* (Engl.: Krebs, A. (1999). *Ethics of Nature.* New York/ Berlin: de Gruyter). Frankfurt: Suhrkamp

Krohn, W. (1987). *Francis Bacon.* München: Beck.

Krohn, W. (2010). Interdisciplinary Cases and Disciplinary Knowledge. In: Frodeman, R. et al. (eds.). *The Oxford Handbook of Interdisciplinarity.* Oxford/ New York: Oxford University Press, pp. 32–49.

Krohn, W., Grunwald, A., Ukowitz, M. (2017). Transdisziplinäre Forschung revisited: Erkenntnisinteressen, Forschungsgegenstände, Wissensform und Methodologie. *Gaia* 26(4), pp. 341–347.

Krohn, W., Küppers, G. (eds.) (1990). *Selbstorganisation. Aspekte einer wissenschaftlichen Revolution.* Braunschweig/Wiesbaden: Vieweg.

Kuhn, T.S. (2012). *The Structure of Scientific Revolutions: 50th anniversary* edition (4th ed.). Chicago: University of Chicago Press.

Kuhlmann, W. (1994). Prinzip Verantwortung versus Diskursethik. In: Böhler, D. (ed.). *Ethik für die Zukunft. Im Diskurs mit Hans Jonas.* München: Beck, pp. 277–302.

Küppers, B.-O. (1993). Chaos und Geschichte. Lässt sich das Weltgeschehen in Formeln fassen? In: Breuer, R. (ed.). *Der Flügelschlag des Schmetterlings.* Stuttgart: DVA, pp. 69–95.

Küppers, B.-O. (2000). Die Strukturwissenschaften als Bindeglied zwischen Natur- und Geisteswissenschaften. In: Küppers, B.-O. (ed.). *Die Einheit der Wirklichkeit. Zum Wissenschaftsverständnis der Gegenwart.* München: Fink, pp. 89–110.

Kwa, C. (2005). Interdisciplinarity and Postmodernity in the Environmental Sciences. *History and Technology* 21, pp. 331–344.

Latucca, L.R. (2001). *Creating Interdisciplinarity: Interdisciplinary Research and Teaching among College and University Faculty.* Nashville, TN: Vanderbilt University Press.

Langer, J.S. (1980). Instabilities and Pattern Formation. *Reviews of Modern Physics* 52, pp. 1–28.

Latour, B. (1987). *Science in Action. How to Follow Scientists and Engineers through Society.* Cambridge, MA: Harvard University Press.

Latour, B. (1990). The Force and the Reason of Experiment. In: Grand, H.E. (ed.). *Experimental Inquiries.* Dordrecht: Kluwer, pp 49–80.

Latour, B. (1993). *We Have Never Been Modern.* Cambridge, MA: Harvard University Press.

Latour, B. (1999). *Pandora's Hope: Essays on the Reality of Science Studies.* Cambridge, MA: Harvard University Press.

Latour, B., Woolgar, S. (1979). *Laboratory Life.* Princeton, NJ: Princeton University Press.

Lenhard, J. (2007). Computer Simulations: The Cooperation between Experimenting and Modeling. *Philosophy of Science* 74, pp. 176–194.

Lenk, H., Ropohl, G. (1987). *Technik und Ethik*. Stuttgart: Reclam.

Leopold, A. (1949). *A Sand County Almanac: And Sketches Here and There*. New York: Oxford University Press.

Levy, D.J. (2002). *Hans Jonas. The Integrity of Thinking*. Columbia/London: University of Missouri Press.

Li, T.-Y., Yorke, J.A. (1975). Period Three Implies Chaos. *The American Mathematical Monthly* 82 (10), pp. 985–992.

Liebert, W., Schmidt, J.C. (2010). Towards a Prospective Technology Assessment. Challenges for Technology Assessment (TA) in the Age of Technoscience. *Praxis & Poiesis* 7(1–2), pp. 99–116.

Lingner, S. (2015). Interdisciplinary Integration in Technology Assessment. A Report from Practise. In: Scherz, C., Michalek, T., Hennen, L., Hebáková, L., Hahn, J., Seitz, J.B. (eds.). *The Next Horizon of Technology Assessment*. Prague: Technology Centre, pp. 359–364.

Lorenz, E.N. (1963). Deterministic Nonperiodic Flow. *Journal of the Atmospheric Sciences* 20, pp. 130–141.

Lorenz, E.N. (1989). Computational Chaos – A Prelude to Computational Instability. *Physica D* 35, pp. 299–317.

Lorenzen, P. (1974). *Konstruktive Wissenschaftstheorie*. Frankfurt: Suhrkamp.

Löwy, I. (1992). The Strength of Loose Concepts. Boundary Concepts, Federative Experimental Strategies and Disciplinary Growth: The Case of Immunology. *History of Science* 30(4), pp. 371–396.

Luhmann, N. (1998). *Die Gesellschaft der Gesellschaft*. Frankfurt: Suhrkamp.

Luhmann, N. (2003). *Soziologie des Risikos*. Berlin/New York: De Gruyter.

Luisi, P.L., Stano, P. (2011). Minimal Cell Mimicry. *Nature Chemistry* 3, pp. 755–756.

Maasen, S. (2010). Transdisziplinarität. Dekonstruktion eines Programms zur Demokratisierung der Wissenschaft. In: Bogner, A., Kastenhofer, K., Torgersen, H. (eds.). *Inter- und Transdisziplinarität im Wandel? Neue Perspektiven auf problemorientierte Forschung und Politikberatung*. Baden-Baden: Nomos, pp. 247–267.

Mach, E. (1988). *Die Mechanik in ihrer Entwicklung (1833)*. Leipzig/Darmstadt: Akademie Verlag.

Machamer, P., Wolters, G. (eds.) (2004). *Science, Values, and Objectivity*. Pittsburgh, PA: University of Pittsburgh Press.

Maier, W., Zoglauer, T. (eds.) (1994). *Technomorphe Organismuskonzepte. Modellübertragungen zwischen Biologie und Technik*. Stuttgart: Frommann Holzboog.

Mainzer, K. (1996). *Thinking in Complexity. The Complex Dynamics of Matter, Mind, and Mankind*. Heidelberg: Springer.

Mainzer, K. (2005). *Symmetry and Complexity. The Spirit and Beauty of Nonlinear Science*. Singapore: World Scientific.

Mandelbrot, B. (1982). *The Fractal Geometry of Nature*. San Francisco: Freeman.

Manson, N.A. (2002). *Formulating the Precautionary Principle. Environmental Ethics* 24(3), pp. 263–274.

Mantegna, R.N., Stanley, H.E. (2000). *An Introduction to Econophysics: Correlations and Complexity in Finance*. Cambridge, UK: Cambridge University Press.

Marliere, P. (2009). The Farther, the Safer: A Manifesto for Securely Navigating Synthetic Species Away from the Old Living World. *Systems and Synthetic Biology* 3, pp. 77–84.

Maxwell, J.C. (1873). Does the Progress of Physical Science Tend to Give Any Advantage to the Opinion of Necessity (or Determinism) Over that of the Contingency of Events and the Freedom of the Will?. In: Campbell, L., Garnett, W. (eds.). *The Life of James Clerk Maxwell (1822)*. New York: Johnson, pp. 434–444.

Maxwell, J.C. (1991). *Matter and Motion (1876)*. New York: Dover.

McCauley, J.L. (2004). *Dynamics of Markets: Econophysics and Finance*. Cambridge, UK: Cambridge University Press.

Meinhardt, H. (1995). *The Algorithmic Beauty of Sea Shells*. Berlin: Springer.

Merton, R.K. (1973). *The Sociology of Science. Theoretical and Empirical Investigations*. Chicago/London: University of Chicago Press.

Mikosch, G. (1993). Interdisziplinarität als kritische Anfrage der Geistes- und Sozialwissenschaften an die naturwissenschaftlichen und technischen Disziplinen oder Auf dem Weg zu einer kritischen Theorie der Interdisziplinarität. In: Arber, W. (ed.). *Inter- und Transdisziplinarität. Warum? Wie?* Bern/Stuttgart/Wien: Haupt, pp. 55–67.

Miller, S., Selgelid, M. (2006). *Ethical and Philosophical Consideration of the Dual-Use Dilemma in the Biological Sciences*. Canberra, Australia: Australian National University/Centre for Applied Philosophy and Public Ethics.

Mitchell, S.D. (2009). *Unsimple Truths. Science, Complexity and Policy*. Chicago: University of Chicago Press.

Mittelstraß, J. (1987). Die Stunde der Interdisziplinarität? In: Kocka, J. (ed.). *Interdisziplinarität. Praxis, Herausforderung, Ideologie*. Frankfurt: Suhrkamp, pp. 152–158.

Mittelstraß, J. (1992). *Leonardo-Welt. Über Wissenschaft, Forschung und Verantwortung*. Frankfurt: Suhrkamp.

Mittelstraß, J. (1998). *Die Häuser des Wissens. Wissenschaftstheoretische Studien*. Frankfurt: Suhrkamp.

Moran, J. (2010). *Interdisciplinarity* (2nd ed.). London: Routledge.

Morgan, M.S., Morrison, M. (eds.) (1999). *Models as Mediators. Perspectives on Natural and Social Science*. Cambridge, UK: Cambridge University Press.

Morozov, E. (2013). *To Save Everything, Click Here. The Folly of Technological Solutionism*. New York: Public Affairs.

Morris, J. (ed.) (2000). *Rethinking Risk and the Precautionary Principle*. Woburn, MA: Butterworth-Heinemann.

Morris, T. (2013). *Hans Jonas's Ethics of Responsibility. From Ontology to Ecology*. New York: SUNY Press.

Muntersbjorn, M.M. (2002). Francis Bacon's Philosophy of Science: Machina intellectus and forma indita. *Philosophy of Science* 70, pp. 1137–1148.

Nachtigall, W. (1994). *Erfinderin Natur*. Hamburg: Rasch & Röhrig.

Næss, A. (1973). Shallow and the Deep, Long-Range Ecology Movement. *Inquiry* 16(1), pp. 95–100.

National Academy of Science. (2005). *Facilitating Interdisciplinary Research*. Washington, DC: The National Academies Press.

Nersessian, N.J. (2012). Engineering Concepts: The Interplay between Concept Formation and Modeling Practices in Bioengineering Sciences. *Mind, Culture, and Activity* 19(3), pp. 222–239.

Nersessian, N., Patton, C. (2009). Model-Based Reasoning in Interdisciplinary Engineering. In: Meijers, A. (ed.). *Handbook of the Philosophy of Technology and Engineering Sciences*. Amsterdam: Elsevier, pp. 687–718.

Newell, W.H. (2001). A Theory of Interdisciplinary Studies. *Issues in Integrative Studies* 19, pp. 1–25.

Newell, W.H. (2013). The State of the Field. Interdisciplinary Theory. *Issues in Interdisciplinary Studies* 31, pp. 22–43.

Newton, I. (1983). *Optik oder Abhandlung über Spiegelungen, Brechungen, Beugungen und Farben des Lichts* (1704/1717). Wiesbaden/Braunschweig: Vieweg.

Nicolescu, B. (2002). *Manifesto of Transdisciplinarity*. Albany, NY: State University of New York Press.

Nicolescu, B. (2008). In Vitro and In Vivo Knowledge. In: Nicolescu, B. (ed.). *Transdisciplinarity. Theory and Practice*. Cresskill, NJ: Hampton Press, pp. 1–22.

Nicolis, G., Prigogine, I. (1977). *Self-Organization in Nonequilibrium Systems. From Dissipative Structures to Order through Fluctuations*. New York/London: Wiley.

Nietzsche, F. (1930). *Die fröhliche Wissenschaft* (1882/1887). Leipzig: Kroener.

Nolfi, S., Floreano, D. (2000). *Evolutionary Robotics: The Biology, Intelligence, and Technology of Self-Organizing Machines*. Cambridge, MA: MIT Press.

Nordmann, A. (2005). Was ist TechnoWissenschaft? In: Rossmann, T., Tropea, C. (eds.). *Bionik: Aktuelle Forschungsergebnisse aus Natur-, Ingenieur- und Geisteswissenschaften*. Berlin: Springer, pp. 291–311.

Nordmann, A. (2006). Collapse of Distance: Epistemic Strategies of Science and Technoscience. *Danish Yearbook of Philosophy* 41, pp. 7–34.

Nordmann, A. (2008a). Philosophy of NanoTechnoScience. In: Fuchs, H.H. et al. (eds.). *Nanotechnology*. Weinheim: Wiley, pp. 217–244.

Nordmann, A. (2008b). Technology Naturalized. A Challenge to Design for the Human Scale. In: Vermaas, P.E. et al. (eds.). *Philosophy and Design*. Heidelberg/New York: Springer, pp. 173–184.

Nordmann, A. (2011). The Age of Technoscience. In: Nordmann, A., Radder, H., Schiemann, G. (eds.). *Science Transformed? Debating Claims of an Epochal Break*. Pittsburgh, PA: University of Pittsburgh Press, pp. 19–30.

Nordmann, A. et al. (2004). *Converging Technologies. Shaping the Future of European Societies. High Level Expert Group 'Foresighting the New Technology Wave'*. Brussels: European Commission.

Nordmann, A., Radder, H., Schiemann, G. (eds.) (2011). *Science Transformed? Debating Claims of an Epochal Break*. Pittsburgh, PA: University of Pittsburgh Press.

Norton, B.G. (2005). *Sustainability. A Philosophy of Adaptive Ecosystem Management*. Chicago: University of Chicago Press.

Norton, B.G. (2015). *Sustainable Values, Sustainable Change. A Guide to Environmental Decision Making*. Chicago: University of Chicago Press.

Nowotny, H., Scott, P., Gibbons, M. (2001). *Re-Thinking Science. Knowledge and the Public in an Age of Uncertainty*. Cambridge, UK: Polity Press.

Omodeo, P.D. (2019). *Political Epistemology. The Problem of Ideology in Science Studies*. New York: Springer.

O'Rourke, M., Crowley, S.J. (2013). Philosophical Intervention and Cross-Disciplinary Science: The Story of the Toolbox Project. *Synthese* 190, pp. 1937–1954.

OTA (1972). *Office of Technology Assessment Act* (Public Law 92-484, 92d Congress, H.R. 10243, October 13, 1972). Washington, DC: OTA Publications.

Ott, K. (1993). *Ökologie und Ethik*. Tübingen: Attempto.

Parker, T.S., Chua, L.O. (1989). *Practical Numerical Algorithms for Chaotic Systems*. New York: Springer.

Passmore, J. (1974). *Man's Responsibility for Nature. Ecological Problems and Western Traditions.* London: Duckworth.

Pauli, W. (1961). *Aufsätze und Vorträge über Physik und Erkenntnistheorie.* Braunschweig: Vieweg.

Peitgen, H.-O., Jürgens, H., Saupe, D. (1998). *Chaos. Bausteine der Ordnung (1994).* Reinbek, Germany: Rowohlt.

Picht, G. (1969). *Wahrheit, Vernunft, Verantwortung: philosophische Studien.* Stuttgart: Klett-Cotta.

Pickering, A. (1984). *Constructing Quarks. A Sociological History of Particle Physics.* Chicago: University of Chicago Press.

Pohl, C., Hirsch Hadorn, G. (2006). *Gestaltungsprinzipien für die transdisziplinäre Forschung.* München: Oekom.

Pohl, C., Hirsch Hadorn, G. (2007). *Principles for Designing Transdisciplinary Research–Proposed by the Swiss Academies of Arts and Sciences.* München: Oekom.

Poincaré, H. (1892). *Les méthodes nouvelles de la mécanique celeste (series 1/2/3).* Paris: Gauthier-Villars.

Poincaré, H. (1914). *Wissenschaft und Methode (1908).* Leipzig: Teubner.

Popper, K.R. (1992). *The Logic of Scientific Discovery (1935).* London/New York: Routledge.

Popper, K.R. (2000). *Vermutungen und Widerlegungen.* Tübingen: Mohr Siebeck.

Pottage, A., Sherman, B. (2007). Organisms and Manufactures: On the History of Plant Inventions. *Melbourne University Law Review* 31, pp. 539–568.

Poser, H. (2001). *Wissenschaftstheorie.* Stuttgart: Reclam.

Prigogine, I. (1980). *From Being to Becoming. Time and Complexity in the Physical Sciences.* New York: Freeman.

Prigogine, I., Glansdorff, P. (1971). *Thermodynamic Theory of Structure, Stability and Fluctuation.* New York/London: Wiley.

Prigogine, I., Stengers, I. (1984). *Order Out of Chaos: Man's New Dialogue with Nature.* New York/London: Bantam Books.

Psillos, S. (1999). *Scientific Realism. How Science tracks Truth.* London/New York: Routledge.

Pühler, A. et al. (eds.) (2011). *Synthetische Biologie. Die Geburt einer neuen Technikwissenschaft.* Berlin/Heidelberg: Springer.

Putnam, H. (2002). *Collapse of the Fact/Value Dichotomy.* Cambridge, MA: Harvard University Press.

Radder, H. (ed.) (2003). *The Philosophy of Scientific Experimentation.* Pittsburgh, PA: Pittsburgh University Press.

Repko, A., Szostak, R. (2017). *Interdisciplinary Research: Process and Theory* (3rd ed.). Thousand Oaks, CA: Sage.

Repko, A., Szostak, R., Buchberger, M. (2017). *Introduction to Interdisciplinary Studies* (2nd ed.), Thousand Oaks, CA: Sage.

Riesch, H., Potter, C. (2014). *Citizen Science as Seen by Scientists: Methodological, Epistemological and Ethical Dimensions. Public Understanding of Science* 23(1), pp. 107–120.

Rittel, H.W., Webber, M.M. (1973). Dilemmas in a General Theory of Planning. *Policy Sciences* 4, pp. 155–169.

Roco, M.C., Bainbridge, W.S. (eds.) (2002). *Converging Technologies for Improving Human Performance. Nanotechnology, Biotechnology, Information Technology and Cognitive Science.* Arlington, VA: National Science Foundation.

Rolston, H. (1988). *Environmental Ethics. Duties to and Values in the Natural World*. Philadelphia: Temple University Press.

Ropohl, G. (2005). Allgemeine Systemtheorie als transdisziplinäre Integrationsmethode. *Technikfolgenabschätzung, Theorie und Praxis* 14(2), pp. 24–31.

Roukes, M.L. (2001). 'Unten gibt's noch viel Platz.' *Spektrum der Wissenschaft Spezial (Nanotechnologie)* 2, pp. 32–39.

Rouse, J. (1987). *Knowledge and Power: Toward a Political Philosophy of Science*. Ithaca, NY: Cornell University Press.

Royal Academy of Engineering. (2009). *Synthetic Biology: Scope, Applications and Implications*. London: Royal Academy of Engineering.

Rueger, A., Sharp, W. D. (1996). Simple Theories of a Messy World: Truth and Explanatory Power in Nonlinear Dynamics. *British Journal for the Philosophy of Science* 47, pp. 93–112.

Ruelle, D., Takens, F. (1971). On the Nature of Turbulence. *Communications in Mathematical Physics* 20, pp. 167–192.

Sandin, P. (1999). *Dimensions of the Precautionary Principle. Human and Ecological Risk Assessment* 5(5), pp. 889–907.

Salmon, W. (1989). Four Decades of Scientific Explanation. In: Kitcher, P., Salmon, W. (eds.). *Scientific Explanation*. Minneapolis, MN: University of Minnesota Press, pp. 3–219.

Salter, L., Hearn, A. (eds.) (1996). *Outside the Lines: Issues in Interdisciplinary Research*. Montreal: McGill-Queens University Press.

Schäfer, L. (1993). Bacon's Project: Should It Be Given Up? *Man and World* 26, pp. 303–317.

Schäfer, L. (1999). *Das Bacon-Projekt*. Frankfurt: Suhrkamp.

Schelsky, H. (1961). Einsamkeit und Freiheit. Zur sozialen Idee der deutschen Universität. *Kölner Zeitschrift für Soziologie und Sozialpsychologie* 13, pp. 702–704.

Schmidt, J.C. (2002). Vom Leben zur Technik? Kultur- und wissenschaftsphilosophische Aspekte der Natur-Nachahmungsthese in der Bionik. *Dialektik. Zeitschrift für Kulturphilosophie* 2, pp. 129–142.

Schmidt, J.C. (2003). Wundstelle der Wissenschaft. Wege durch den Dschungel der Interdisziplinarität. *Scheidewege* 33, pp. 169–189.

Schmidt, J.C. (2004). Unbounded Technologies. Working through the Technological Reductionism of Nanotechnology. In: Baird, D. et al. (eds.). *Discovering the nanoscale*. Amsterdam: AOS, pp. 35–51.

Schmidt, J.C. (2005). Dimensionen der Interdisziplinarität. Wege zu einer Wissenschaftstheorie der Interdisziplinarität. *Technikfolgenabschätzung. Theorie und Praxis* 14(2), pp. 12–17.

Schmidt, J.C. (2007a). Die Aktualität der Ethik von Hans Jonas. Eine Kritik der Kritik des Prinzips Verantwortung. *Deutsche Zeitschrift für Philosophie* 4, pp. 545–569.

Schmidt, J.C. (2007b). Knowledge Politics of Interdisciplinarity. Specifying the Type of Interdisciplinarity in the NSF's NBIC Scenario. *Innovation, European Journal of Social Science Research* 20(4), pp. 313–328.

Schmidt, J.C. (2008a). *Instabilität in Natur und Wissenschaft. Eine Wissenschaftsphilosophie der nachmodernen Physik*. Berlin: De Gruyter.

Schmidt, J.C. (2008b). Towards a Philosophy of Interdisciplinarity. An Attempt to provide a Classification and Clarification. *Poiesis & Praxis* 5(1), pp. 53–69.

Schmidt, J.C. (2008c). From Symmetry to Complexity. On Instabilities and the Unity in Diversity in Nonlinear Science. *International Journal of Bifurcation and Chaos* 18(4), pp. 897–910.

Schmidt, J.C. (2011a). Challenged by Instability and Complexity. On the Methodological Discussion of Mathematical Models in Nonlinear Sciences and Complexity Theory. In: Hooker, C. (ed.). *Philosophy of Complex Systems.* Amsterdam/Boston: Elsevier, pp. 223–254.

Schmidt, J.C. (2011b). The Renaissance of Francis Bacon. On Bacon's Account of recent Nanotechnosciences. *NanoEthics* 5(1), pp. 29–41.

Schmidt, J.C. (2013). Defending Hans Jonas' Environmental Ethics: On the Relation between Philosophy of Nature and Ethics. *Environmental Ethics* 35(4), pp. 461–479.

Schmidt, J.C. (2015a). *Das Andere der Natur. Neue Wege der Naturphilosophie.* Stuttgart: Hirzel.

Schmidt, J.C. (2015b). Synthetic Biology as Late-Modern Technology. Inquiring into the Rhetoric and Reality of a New Technoscientific Wave. In: Giese, B., Pade, C., Wigger, H., Gleich, A.v. (eds.). *Synthetic Biology. Character and Impact.* Heidelberg/New York: Springer, pp. 1–30.

Schmidt, J.C. (2016). Towards a Prospective Technology Assessment of Synthetic Biology. Fundamental and Propaedeutic Reflections in order to enable an Early Assessment. *Science and Engineering Ethics* 22(4), pp. 1151–1170.

Schmidt, J.C. (2017). Science in an Unstable World. On Pierre Duhem's Challenge to the Methodology of Exact Sciences; In: Pietsch, W., Wernecke, J., Ott, M. (eds.). *Berechenbarkeit der Welt? Philosophie und Wissenschaft im Zeitalter von Big Data.* Wiesbaden: Springer VS, pp. 403–434.

Schmidt, J.C. (2019). Is There Anything New Under the Sun? An Approach from the Philosophy of Science. In: Lüttge, U., Wegner, L. (eds.). *Emergence and Modularity in Life Sciences and Beyond.* Berlin/New York: Springer, pp. 3–36.

Schmidt, M. (2009). Do I Understand What I Can Create? Biosafety Issues in Synthetic Biology. In: Schmidt, M. et al. (eds.). *Synthetic Biology. The Technoscience and Its Societal Consequences.* Berlin: Springer, pp. 81–100.

Schmidt, M., de Lorenzo, V. (2012). Synthetic Constructs in/for the Environment: Managing the Interplay between Natural and Engineered Biology. *Federation of the European Biochemical Societies* 586, pp. 2199–2206.

Schneidewind, U., Singer-Brodowski, M. (2013). *Transformative Wissenschaft. Klimawandel im deutschen Wissenschafts- und Hochschulsystem.* Marburg, Germany: Metropolis.

Scholz, R.W. (2011). *Environmental Literacy in Science and Society: From Knowledge to Decisions.* Cambridge, UK: Cambridge University Press.

Schot, J., Rip, A. (1996). The Past and Future of Constructive Technology Assessment. *Technological Forecasting and Social Change* 54, pp. 251–268.

Schulz, W. (1972). *Philosophie in der veränderten Welt.* Pfullingen, Germany: Neske.

Schummer, J. (2004). Interdisciplinary Issues of Nanoscale Research. In: Baird, D. et al. (eds.). *Discovering the Nanoscale.* Amsterdam: IOS, pp. 9–20.

Schurz, R. (1995). Ist Interdisziplinarität möglich? *Universitas* 11, pp. 1080–1089.

Schweitzer, A. (1966). *The Teaching of Reverence for Life.* London: Owen.

Schwille, P. (2011). Bottom-Up Synthetic Biology: Engineering in a Tinkerer's World. *Science* 333, pp. 1252–1254.

Schwille, P., Diez, S. (2009). Synthetic Biology of Minimal Systems. *Critical Reviews in Biochemistry and Molecular Biology* 44(4), pp. 223–242.

Segerstrale, U. (ed.) (2000). *Beyond Science Wars: The Missing Discourse about Science and Society*, New York: State University of New York Press.

Serres, M. (1992). *Hermes II: Interferenz*. Berlin: Merve.

Sismondo, S. (2005). *An Introduction to Science and Technology Studies*. Malden, MA: Blackwell.

Smalley, R.E. (2001). Chemie, Liebe und dicke Finger. *Spektrum der Wissenschaft Spezial (Nanotechnologie)* 2, pp. 66–67.

Snow, C.P. (2001). *The Two Cultures (1959)*. London: Cambridge University Press.

Sokal, A. (1996). Transgressing the Boundaries: Towards a Transformative Hermeneutics of Quantum Gravity. *Social Text* 6(47), pp. 217–252.

Sokal, A., Bricmont, J. (1998). *Fashionable Nonsense: Postmodern Intellectuals' Abuse of Science*. New York: Picador.

Solla Price, D.J. de (1963). *Little Science, Big Science*. New York: Columbia University Press.

Star, S.L., Griesemer, J.R. (1989). Institutional Ecology: "Translations" and Boundary Objects: Amateurs and Professionals in Berkeley's Museum of Vertebrate Zoology 1907–39. *Social Studies of Science* 19, pp. 387–420.

Stehr, N. (2005). *Knowledge Politics: Governing the Consequences of Science and Technology*. Boulder, CO: Paradigm.

Strohschneider, P. (2014). Zur Politik der Transformativen Wissenschaft. In: Brodocz, A., Herrmann, D., Schmidt, R., Schulz, D. Schulze-Wessel, J. (eds.). *Die Verfassung des Politischen. Festschrift für Hans Vorländer*. Wiesbaden: Springer, pp. 175–192.

Szostak, R. (2015). Extensional Definition of Interdisciplinarity. *Issues in Interdisciplinary Studies* 33, pp. 94–117.

Szostak, R., Gnoli, C., Lopez-Huertas, M. (2016). *Interdisciplinary Knowledge Organization*. Berlin: Springer.

Taylor, P. W. (1986). *Respect for Nature: A Theory of Environmental Ethics*. Princeton, NJ: Princeton University Press.

TESSY (2008). *Information Leaflet: Synthetic Biology in Europe*. http://www.tessy-europe.eu/public_docs/SyntheticBiology_TESSY-Information-Leaflet.pdf. Accessed 23 May 2020.

Thom, R. (1975). *Structural Stability and Morphogenesis. An Outline of a General Theory of Models* (1972). Reading, MA: Benjamin.

Tucker, J.B., Zilinskas, R.A. (2006). The Promise and Perlis of Synthetic Biology. *New Atlantis* 12, pp. 25–45.

Turner, S. (2000). What Are Disciplines? And How Is Interdisciplinarity Different? In: Weingart, P., Stehr, N. (eds.). *Practising Interdisciplinarity*. Toronto: University of Toronto Press, pp. 46–65.

Venter, C. (2010). The Creation of 'Synthia'. Synthetic Life. *Science Interview*. 23 May 2010, The Naked Scientists. http://www.thenakedscientists.com/HTML/content/interviews/interview/1332/. Accessed 20 June 2019.

Vriend, H. d. (2006). *Constructing Life: Early Social Reflections on the Emerging Field of Synthetic Biology (Working Document 9)*. The Hague: Rathenau Institute.

Vuillemin, J. (1991). Introduction. In: Duhem, P. (ed.). *The Aim and Structure of Physical Theory* (1906). Princeton, NJ: Princeton University Press, pp. xv–xxxiii.

WCED (World Commission on Environment and Development). (1987). *Our Common Future*. Oxford: Oxford University Press/New York: United Nation Press.

Weber, J. (2003). *Umkämpfte Bedeutungen. Naturkonzepte im Zeitalter der Technoscience*. Frankfurt: Campus.

Wechsler, D., Hurst, A.C. (2011). Interdisziplinäre Systemintegration und Innovationsgenese: Ein methodologischer Zugang für die interdisziplinäre Forschung. *Journal for the General Philosophy of Science* 42, pp. 141–155.

Weinberg, A.M. (1972). Science and Trans-Science. *Minerva* 10(2), pp. 209–222.

Weingart, P. (2000). Interdisciplinarity – The Paradoxical Discourse. In: Weingart, P., Stehr, N. (eds.) (2000). *Practising Interdisciplinarity*. Toronto: University of Toronto Press, pp. 25–41.

Weingart, P., Stehr, N. (eds.) (2000). *Practising Interdisciplinarity*. Toronto: University of Toronto Press.

Weischedel, W. (1958). *Das Wesen der Verantwortung*. Frankfurt: Klostermann.

Weizsäcker, C.F.v. (1974). *Die Einheit der Natur*. München: dtv.

Wetz, F.J. (1994). *Hans Jonas zur Einführung*. Hamburg: Junius Verlag.

Wickson, F., Carew, A. L., Russell, A. (2006). Transdisciplinary Research: Characteristics, Quandaries and Quality. *Futures* 38(9), pp. 1046–1059.

Wiener, N. (1968). *Kybernetik. Regelung und Nachrichtenübertragung in Lebewesen und Maschine*. Reinbek, Germany: Rowohlt.

Wiggins, S. (1988). *Global Bifurcations and Chaos. Analytical Methods*. New York: Springer.

Wilholt, T. (2012). *Die Freiheit der Forschung*. Frankfurt: Suhrkamp.

Wilson, E.O. (1998). *Consilience. The Unity of Knowledge*. London: Little, Brown.

Winner, L. (1980). Do Artifacts Have Politics? *Daedalus* 109(1), pp. 121–136 (reprinted in: Teich, A. (ed.) (2006). *Technology and the Future*. Belmont, CA: Thomson/Wadsworth, pp. 50–66.)

Wolters, G. (2004). Problem. In: *Enzyklopädie für Philosophie und Wissenschaftstheorie* (ed. Mittelstraß, J., Bd. 3). Stuttgart: Metzler, pp. 334–347.

Woodward, J. (2000). Explanation and Invariance in the Special Sciences. *British Journal for the Philosophy of Science* 51, pp. 197–254.

Wright, G.H.V. (1971). *Explanation and Understanding*. London: Routledge.

Ziman, J. (1987). *Knowing Everything about Nothing. Specialization and Change in Scientific Careers*. Cambridge, UK: Cambridge University Press.

Ziman, J. (2000). Postacademic Science: Constructing Knowledge with Networks and Norms. In: Segerstrale, U. (ed.). *Beyond Science Wars: The Missing Discourse about Science and Society*. New York: State University of New York Press, pp. 135–154.

Zittel, C., Engel, G., Nanni, R., Karafyllis, N.C. (eds.) (2008). *Philosophies of Technology: Francis Bacon and His Contemporaries*. Leiden, The Netherlands: Brill.

Zweck, A. (2002). Technologiefrüherkennung. Ein Instrument der Innovationsförderung. *Wissenschaftsmanagement. Zeitschrift für Innovation* 2, pp. 25–30.

Index

The use of 'f' in the index indicates that the term also appears within the next one or two pages.

Printed in the United States
by Baker & Taylor Publisher Services

Printed in the United States
by Baker & Taylor Publisher Services